Modelling Non-Linear Wave Processes

Titles of related interest from VNU Science Press

Journals

Journal of Computational Mathematics
Soviet Journal of Numerical Analysis and Mathematical Modelling

Books

BAHTURIN: Identical Relations in Lie Algebra
ROMANOV: Inverse Problems of Mathematical Physics
LEVITAN: Inverse Sturm-Louiville Problems

Modelling Non-Linear Wave Processes

Yu. A. Berezin

Institute of Theoretical and Applied Mechanics,
Siberian Branch of the USSR Academy of Sciences

Translated by L. Ya. Yuzina

XW VNU SCIENCE PRESS///   Utrecht, The Netherlands

1987

VNU Science Press BV
P.O. Box 2093
3500 GB Utrecht
The Netherlands

© 1987 VNU Science Press BV

First English edition published 1987

ISBN 90-6764-075-1

CIP-DATA KONINKLIJKE BIBLIOTHEEK, DEN HAAG

Modelling non-linear wave processes/Ya. A. Berezin; transl. by L. Ya Yuzina. — Utrecht: VNU Science Press. — Ill. With refs, Ill.
SISO 533 UDC 533.9
Subject heading: non-linear wave processes.

Printed in DDR by Druckhaus Köthen

Foreword

This monograph is devoted to analytical and numerical research of non-linear wave processes in dispersive and unstable media. Mathematical models and computation algorithms are presented in detail.

This book will be useful for research workers, post-graduate students and specialists in numerical methods of solving problems in mathematical physics.

Academician N. N. Yanenko

Contents

Contents

Introduction

Wave processes are widely spread in the world and, hence, practically all branches of physics have to deal with their investigation — both for gaining insight into the fundamental laws of nature and for employing them in important practical applications. Due to advances in plasma research and the creation of powerful acoustic and light sources, much attention has also been given to non-linear wave processes. Such situations are, as a rule, described by complex non-linear equations in partial derivatives. Therefore, together with conventional analytical methods of theoretical investigation (similarity solutions, asymptotic expansions, subtle non-standard substitutions), there exists a universal and powerful tool for solving problems such as numerical methods, allowing one to computerize not only single computations but also their whole series — the so-called numerical experiments for determining basic features of the processes studied.

This monograph is devoted to numerical investigations of non-linear wave processes in various media, with a detailed description of the corresponding models, numerical methods, algorithms and physical results. Basically, solutions to problems, in which the author took part are considered.

Chapter 1 deals with some dispersive media and linear wave processes, amongst them: gravitational and gravitation ripple waves on a fluid surface in a channel of a finite depth, small-amplitude waves in a fluid with gas bubbles, in blood vessels, in superfluid helium, in plasma with no magnetic field, in cold plasma with a magnetic field. Unstable media are considered, wherein small disturbances in the parameters increase with growing time in a limited spectrum domain, i.e. anisotropic rarefied plasma, viscous films on solid surfaces.

Chapter 2 concentrates on waves of a finite amplitude in weak dispersive and unstable media. The evolution of disturbances of a small but finite amplitude in a dispersive medium is described by the Korteweg–de Vries equation (KdV), that in the presence of dissipation — by the Burgers–Korteweg–de Vries equation (BKV). Some simple and effective numerical algorithms to solve these equations are presented. One-dimensional Alfven waves of a finite amplitude, propagating along a magnetic field in a rarefied anisotropic plasma, have been studied both analytically and numerically. The system of equations by Chew, Goldberger and Low, describing plasma with the finite Larmor ion radius taken into account, has been demonstrated to have analytical solutions in the form of circularly polarized monochromatic waves with their amplitudes changing periodically in time. This Chapter also presents some results of a numerical solution of the problem of viscous films falling along solid surfaces.

Chapter 3 features non-linear and shock waves in a rarefied plasma, based on gas-dynamic type models. Such models are basically determined by the effects associated with a collective interaction of the electro-magnetic field and plasma particles that are self-consistently introduced into the equations of moments through the effective frequency of collisions and 'turbulent' coefficients of resistivity and heat conductivity. For stationary shock waves, propagating in such a plasma at an arbitrary angle to the undisturbed magnetic field, some critical parameters, at which an isomagnetic jump in density takes place, have been found. The Chapter presents an effective numerical method for solving one-dimensional non-stationary equations of a two-component plasma with account taken of dispersion, anisotropy, anomalous resistivity, heat conductivity, and the results of the numerical solution and a classification of various conditions based on these results. This Chapter gives solutions to the problem of the structure of a stationary compression wave in a three-component plasma composed of electrons and ions of two types, to the problem of expansion of a plasma cylinder against a less dense background (under the supposition of a cylindrical symmetry), as well as to the problem of the two-dimensional bow shock wave arising in a rarefied plasma flowing around a conducting cylinder under the conditions when the plasma is a dispersive medium for ion-acoustic waves.

Chapter 4 explores strong, collisionless shock waves, for which the phenomenon of overturning is possible, on the basis of combined or hybrid models, where the ionic plasma component is described by the Vlasov kinetic equation, while the electronic one — by equations of a gas-dynamic type. The basic algorithmic aspects of the method are presented. As an illustration, the Chapter gives the results of a numerical investigation of unsteady one-dimensional shock waves of an arbitrary amplitude, propagating across the magnetic field in a quasi-neutral collisionless plasma.

Chapter 5 reports the results of a numerical solution of self-consistent problems of two-dimensional single-fluid magnetohydrodynamics, associated with the formation of bow shock waves, arising when a viscous heat-conducting plasma of a finite conductivity flows around blunt bodies, and with the processes of reconnection of the lines of force of the magnetic field in plasma, playing a vital part in some physical phenomena.

Chapter 6 details the gas-dynamic processes occurring in self-gravitating neutron matter with the equations of state of the van der Waals type.

The author is grateful to academician R. Z. Sagdeev for the formulation of a number of problems, to academician Drs. N. N. Yanenko, V. S. Imshennik, V. A. Vshivkov, G. I. Dudnikova, O. E. Dmitrieva and P. V. Khenkin for their cooperation in the discussion of the problems considered.

Yu. A. Berezin

Chapter 1

Linear Processes in Weak Dispersive and Unstable Media

To date both theoretical and experimental studies of the problems related to wave processes in dispersive media have received great consideration and development.

Prior to defining a dispersive medium, let us consider the following phenomenon. The propagation of small-amplitude disturbances in any continuous medium is described by linear equations, obtained by a standard procedure of linearizing the governing non-linear equations. For a plane case a general solution of these linear equations has the form:

$$f(x, t) = (2\pi)^{-1} \int_{-\infty}^{\infty} A(k) \exp i(kx - \omega t)\, dk,$$

where $f(x, t)$ is a function describing a wave process, and $A(k) = \int_{-\infty}^{\infty} f(x, 0) \exp(-ikx)\, dx$ is a Fourier component of the initial disturbance. The condition of the existence of a non-trivial solution results in a relationship between the frequency and the wave number $\omega = \omega(k)$. This relation is termed a dispersive equation or the law of dispersion. The phase and group velocities of the propagation of plane waves with small amplitudes are defined by the formulae $v_{ph} = \omega/k$, $v_g = \partial\omega/\partial k$. Now we can give the following definition: a medium is referred to as dispersive with respect to small-amplitude waves of a certain type, if both the phase and group velocities of these waves depend on the wavelength (or the wave number), i.e. $v_{ph} = v_{ph}(k)$, $v_g = v_g(k)$. In other words, the relationship between the frequency and the wave number is non-linear. In a dispersive medium the group and phase wave velocities do not coincide.

There is a great number of media, exhibiting their dispersive properties with respect to various wave processes. Here are some examples.

Let us consider the so-called gravitational waves on the surface of an incompressible fluid with the density $\varrho = $ const. in a channel of a finite depth H (see, for instance, [71]). With the viscosity neglected, the macroscopic velocity of the fluid obeys the equations

$$v_t + (v\nabla)\, v = -\frac{1}{\varrho}\, \nabla p + g, \quad \text{div } v = 0,$$

where p is the pressure of the fluid, g is the gravitational acceleration. For a potential flow, setting $v = \nabla \varphi$ (φ is the velocity potential), we get $\Delta \varphi = 0$. Now we linearize the equation of motion, going over to the waves of small amplitudes:

$$v_t = -\frac{1}{\varrho}\, \nabla p + g \quad \text{or} \quad \nabla \varphi_t = -\nabla(p/\varrho) + g. \qquad (1.1)$$

Let us direct the z-axis upward, perpendicular to the undisturbed horizontal fluid surface (also perpendicular to the channel bottom), having placed the coordinate $z = 0$ on this surface; on the bottom we have $z = -H$. Let the x-axis be directed along the undisturbed fluid surface. Projecting equation (1.1) on the z-axis and integrating it, we get the Bernoulli integral:

$$\varphi_t + p/\varrho + gz = 0. \qquad (1.2)$$

Using (1.2) we can find a boundary condition on the free fluid surface. Let $\zeta(x, y)$ be the z-coordinate of the surface points; since the coordinate $z = 0$ has been chosen on the undisturbed surface, under equilibrium $\zeta = 0$. At $z = \zeta$ formula (1.2) yields

$$\varphi_t + p_0/\varrho + g\zeta = 0$$

where p_0 is the atmospheric pressure. Introducing a new potential $\varphi' = \varphi + (p_0/\varrho)\, t$ and omitting the prime, we get

$$\varphi_t|_{z=\zeta} + g\zeta = 0. \tag{1.3}$$

On the fluid surface the vertical component equals $v_z = \zeta_t$ or

$$\varphi_z|_{z=\zeta} = \zeta_t. \tag{1.4}$$

Formula (1.3) affords $\zeta = -\dfrac{1}{g}\,\varphi_t|_{z=\zeta}$, therefore, equation (1.4) yields a boundary condition for the velocity potential on a free fluid surface

$$(\varphi_{tt} + g\varphi_z)|_{z=\zeta} = 0. \tag{1.5}$$

Since we consider small-amplitude waves, the boundary condition (1.5) can be transferred to the undisturbed surface $z = 0$. In this case the system of equations and the boundary conditions for small-amplitude gravitational waves will be:

$$\Delta\varphi = 0, \quad (\varphi_{tt} + g\varphi_z)|_{z=0} = 0, \quad \varphi_z|_{z=-H} = 0. \tag{1.6}$$

Assuming independence from the coordinate y, we get from (1.6):

$$\varphi_{xx} + \varphi_{zz} = 0, \quad (\varphi_{tt} + g\varphi_z)|_{z=0} = 0, \quad \varphi_z|_{z=-H} = 0.$$

Looking for a solution of this problem in the form $\varphi(x, z, t) = \bar{\varphi}(z) \cos{(kx - \omega t)}$, we find the law of dispersion for gravitational waves

$$\omega^2 = gk\,th(kH). \tag{1.7}$$

For the limiting case of a 'shallow' water, when $kH \ll 1$, $\lambda \gg H$, from (1.7) we have

$$\omega \approx (gH)^{1/2}\, k\left(1 - \frac{1}{6}\, H^2 k^2\right),$$

$$v_{ph} \approx (gH)^{1/2}\left(1 - \frac{1}{6}\, H^2 k^2\right), \tag{1.8}$$

$$v_g \approx (gH)^{1/2}\left(1 - \frac{1}{2}\, H^2 k^2\right).$$

Thus, 'shallow' water is a dispersive medium for surface gravitational waves: shorter waves (with greater wave numbers) propagate with a lower velocity than longer ones. The dispersion of type (1.8) is termed weak, since $v_{ph} \rightarrow (gH)^{1/2} = $ const. at $k \rightarrow 0$.

With decreasing channel depth, surface tension should be taken into account. For a slightly curved surface the Laplace formula gives that $p = p_0 - \alpha \zeta_{xx}$, where p is the fluid pressure near the surface, p_0 is the atmospheric pressure, α is the coefficient of the surface tension; dependence on the coordinate y is neglected. Employing the procedures analogous to those mentioned above, we get a system of equations and boundary conditions for the surface waves in a channel of a finite depth, with the forces of surface tension accounted for

$$\varphi_{xx} + \varphi_{zz} = 0, \quad (\varrho \varphi_{tt} + \varrho g \varphi_z - \alpha \varphi_{xxz})|_{z=0} = 0, \quad \varphi_z|_{z=-H} = 0,$$

which results in the law of dispersion

$$\omega^2 = gk(1 + \alpha k^2/\varrho g) \, th(kH).$$

If $\alpha k^2/\varrho g \ll 1 (\lambda^2 \gg \alpha/\varrho g)$, we have the gravitational waves studied earlier; if $\alpha k^2/\varrho g \gg 1 (\lambda^2 \ll \alpha/\varrho g)$, we have the so-called ripple waves with the law of dispersion $\omega^2 = (\alpha/\varrho) \, k^3 th(kH)$.

Of interest is an intermediate case of ripple-gravitational long waves on a 'shallow' water surface with the equation of dispersion

$$\omega \approx (gH)^{1/2} k \left[1 - \frac{1}{6} (H^2 - 3\alpha/\varrho g) \, k^2 \right]. \tag{1.9}$$

The formula shows that at $H < (3\alpha/\varrho g)^{1/2}$ a qualitative change in the law of dispersion, as compared to gravitational waves, takes place. In this case shorter waves (with greater wave numbers) propagate with greater velocities than longer ones.

Let us now consider the case of small-amplitude waves in an incompressible fluid (density ϱ_f), filled with gas bubbles (density ϱ_g, a bubble radius R, the number of bubbles per unit volume n) [109]. In an undisturbed state the pressure in the fluid and bubbles is p_0;

besides, let us assume, for the sake of simplicity, that all the bubbles have the same radius R_0. A fluid–gas mixture can be considered a continuous medium, if noticeable changes in velocity and pressure occur at the distances greater than those among the bubbles. Therefore, when studying wave processes with certain characteristic wavelengths λ we assume that: (a) the domains small as compared to λ contain many bubbles, i.e. $n^{-1/3} \ll \lambda$; (b) $R_0 \ll \lambda$. Both the velocity and pressure in the mixture, averaged over the domains of such dimensions, are equal to u and p. The mixture density $\varrho = \varrho_f(1 - nV) + \varrho_g nV$ ($V = \dfrac{4}{3}\pi R^3$ is the volume of a bubble). Since $\varrho_g \ll \varrho_f$, we may assume $\varrho = \varrho_f(1 - nV)$.

For a one-dimensional mixture the equations of continuity and motion can be written in a conventional form:

$$\varrho_t + (\varrho u)_x = 0, \qquad \varrho(u_t + uu_x) = -p_x. \qquad (1.10)$$

A more complex problem is to find an equation of state, relating pressure and density. As noted in [109], an isothermal behaviour of gas bubbles in a fluid has been experimentally confirmed, therefore,

$$p_g/\varrho_g = \text{const.} = p_0/\varrho_{g_0} \qquad (1.11)$$

and, assuming that a gas mass is constant in a unit mixture mass, we get

$$\varrho_g nV/(1 - nV) = \text{const.} = \varrho_{g_0} n_0 V_0/(1 - n_0 V_0). \qquad (1.12)$$

Extension and contraction of a bubble in a fluid, neglecting heat losses and those associated with excitation of acoustic vibrations, are described by the Rayleigh equation

$$\varrho_f\left(R\ddot{R} + \frac{3}{2}\dot{R}^2\right) = p_g - p$$

which affords

$$p = p_g - \varrho_f\left(R\ddot{R} + \frac{3}{2}\dot{R}^2\right). \qquad (1.13)$$

An assumption on a constant mass of the bubble: $m = \varrho_g V = \varrho_g \cdot \dfrac{4}{3} \pi R^3 = \dfrac{4}{3} \pi R_0^3 \varrho_{g_0}$ and formula (1.11) show a relation between the bubble radius and gas pressure

$$R = R_0 (p_g / p_0)^{-1/3}. \tag{1.14}$$

Equations (1.10)–(1.14) are a simplified mathematical model of a mixture of a fluid and gas bubbles. Linearizing this system of equations, we find the following law of dispersion for plane waves of small amplitudes, propagating in such a mixture

$$\omega^2 = c_s^2 k^2 (1 + \bar{\beta} k^2)^{-1}, \quad v_{ph} = c_s (1 + \bar{\beta} k^2)^{-1/2} \tag{1.15}$$

where

$$c_s^2 = p_0 / \varrho_f (1 - n_0 V_0)\, n_0 V_0, \quad \bar{\beta} = R_0^2 / 3(1 - n_0 V_0)\, n_0 V_0.$$

For sufficiently long waves $\bar{\beta} k^2 \ll 1$ we get

$$\omega \approx c_s k \left(1 - \dfrac{1}{2} \bar{\beta} k^2\right), \quad v_{ph} \approx c_s \left(1 - \dfrac{1}{2} \bar{\beta} k^2\right) \tag{1.16}$$

that coincides to the accuracy of notations with the law of dispersion (1.8) for gravitational waves on a 'shallow' water surface. Note, that relations (1.16) hold at $\bar{\beta} k^2 \ll 1$, $\lambda^2 \gg \bar{\beta} \approx R_0^2 / n_0 V_0$, i.e. at a small gas content $n_0 V_0 \ll 1$.

Let us now consider propagation of small-amplitude waves in a fluid in a channel with elastic walls, for instance, a mathematical model of blood flow in blood vessels (gemodynamics). For large arteries, blood can be viewed as an incompressible non-viscous fluid with the density ϱ. Let us assume that a blood vessel is an isotropic thin-walled elastic tube with thickness h and density $\bar{\varrho}$. Since blood vessels function under considerable axial tensions, let us assume that a vessel wall undergoes only radial axis-symmetrical displacements. Then, in the linear approximation, the equation of motion for an elastic vessel wall can be as follows (see, for instance [107, 77]):

$$\bar{\varrho} h \bar{u}_{tt} = p + \bar{\varrho} \, \dfrac{h^3}{12} \bar{u}_{ttzz} - \dfrac{Eh\bar{u}}{(1 - v^2) R^2} - \dfrac{Eh^3}{12(1 - v^2)} \bar{u}_{zzzz} \tag{1.17}$$

where $\bar{u}$ is the radial displacement of the vessel wall, p is the pressure disturbance on the wall, R is the undisturbed radius, E is the Young modulus, v is the Poisson coefficient; the z-axis is directed along the blood vessel. The second term in the right-hand side describes the effect of rotational inertia, $h^3/12$ is the moment of the wall inertia; the third term in the right-hand side presents the compression, or the circumferential stress, $Eh/(1 - v^2)$ is the compression rigidity; the last term is the bend, $Eh^3/12(1 - v^2)$ is the bend rigidity.

Introducing, as in the case of gravitational waves, the velocity potential φ, let us write some necessary in linear approximation relations:

$p = -\varrho\varphi_t|_{r=R}$ is the Bernoulli integral,

$\bar{u}_t = \varphi_r|_{r=R}$ is the kinematic condition on the wall,   (1.18)

$\varphi_r|_{r=0} = 0$ is the condition on the tube axis.

Differentiating equation (1.17) with respect to time and using formulae (1.18), we come to the conclusion that the equation for the velocity potential

$$\frac{1}{r}\frac{\partial}{\partial r}\left(r\frac{\partial\varphi}{\partial r}\right) + \frac{\partial^2\varphi}{\partial z^2} = 0 \qquad (1.19)$$

should be solved with the following boundary conditions:

$$\varphi_r|_{r=0} = 0,$$

$$\left(\varrho\varphi_{tt} + \bar{\varrho}h\varphi_{ttr} + \frac{Eh^3}{12(1 - v^2)}\varphi_{zzzzr}\right.$$

$$\left. + \frac{Eh}{(1 - v^2)R^2}\varphi_r - \frac{\bar{\varrho}h^3}{12}\varphi_{ttzzr}\right)\Big|_{r=R} = 0.$$

A solution to equation (1.19) can be written as $\varphi(r, z, t) = I_0(kr)\cos(kz - \omega t)$, where $I_0(kr)$ is the Bessel function of the zero order of an imaginary argument. Substituting this solution into the above boundary condition at $r = R$, we derive the law of dispersion for the waves propagating in a channel having elastic walls:

$$\omega^2 = \frac{2v_{ph}^2}{R} \cdot \frac{k\left(1 + \frac{1}{12} h^2 R^2 k^4\right) I_1(kR)}{I_0(kR) + (\bar{\varrho}/\varrho)\, kh\, I_1(kR)\left(1 + \frac{1}{12} k^2 h^2\right)}. \quad (1.20)$$

Here $v_{ph}^2 = Eh/2\varrho R(1 - v^2)$, $I_1(kR)$ is the Bessel function of the first order of an imaginary argument. In a long-wave approximation, setting $kR \ll 1$, $kh \ll 1$, equation (1.20) gives

$$\omega \approx v_{ph}k \left[1 - \frac{1}{4}\left(\frac{1}{4} + \bar{\varrho}h/\varrho R\right) R^2 k^2\right] \quad (1.21)$$

which shows that the phase velocity of small oscillations under consideration decreases with growing wave number.

Let us now consider small-amplitude waves in superfluid helium. This, in a physical sense, highly interesting medium has been intensively investigated since the cornerstone works by P. L. Kapitsa and L. D. Landau (1938–1941) were published. In superfluid helium the undamping temperature waves, known as the second sound, can propagate; the possibility of their propagation being a specific property of superfluid helium. At the temperature $T = 2.19\ K \equiv T_\lambda$ (λ-point) a phase transition of the second order takes place, while the superfluid component density and the second sound velocity become zero. At temperatures close to T_λ one should take into account the relaxation phenomena which cause dissipation and dispersion of the temperature waves.

The governing equations are the equations of two-velocity hydrodynamics by Landau–Khalatnikov, which, at small gradients of the required functions, have the form [64]:

$$\frac{\partial \boldsymbol{v}_s}{\partial t} + \nabla \left(\frac{1}{2} v_s^2 + \mu + \mu_s\right) = 0,$$

$$\frac{\partial}{\partial t}(\varrho_s + \varrho_n) + \mathrm{div}\,(\varrho_s \boldsymbol{v}_s + \varrho_n \boldsymbol{v}_n) = 0,$$

$$\frac{\partial}{\partial t}(\varrho_s v_{si} + \varrho_n v_{ni}) + \frac{\partial}{\partial x_k}(\varrho_n v_{ni} v_{nk} + \varrho_s v_{si} v_{sk} + p\delta_{ik}) = 0,$$

$$\frac{\partial s}{\partial t} + \text{div } Sv_n = \frac{2\Lambda m}{\hbar}\left(\mu_s + \frac{(v_n - v_s)^2}{2}\right)^2 \varrho_s, \qquad (1.22)$$

$$\frac{\partial \varrho_s}{\partial t} + \text{div } \varrho_s v_s = -\frac{2\Lambda m}{\hbar}\left(\mu_s + \frac{(v_n - v_s)^2}{2}\right)\varrho_s.$$

Here the index s refers to the superfluid component, the index n to the normal component, μ is the chemical potential, Λ is the dimensionless kinetic potential proportional to the correlation length ξ on which the correlation of the velocity potential phase of the superfluid component damps. The second sound can propagate only when the condition $k\xi \ll 1$ is met, otherwise it will be completely damped. Density relaxation of the superfluid component ϱ_s is described by the right-hand side of the last equation of system (1.22).

In a linear approximation one can easily get the following law of dispersion in a long-wave approximation [81]:

$$\omega \approx c_{20}k - i\bar{\alpha}k^2 - \bar{\beta}k^3 \qquad (1.23)$$

where $\bar{\alpha} = \frac{1}{2}(c_{2\infty}^2 - c_{20}^2)\tau$, $\bar{\beta} = \frac{1}{2}(c_{2\infty}^2 - c_{20}^2)c_{20}\tau^2$, c_{20} is the equilibrium velocity of the second sound at $\omega \to 0$, $c_{2\infty}$ is the second sound velocity in the limit $\omega\tau \gg 1$, when the value of the density of the equilibrium component ϱ_s is not in time to get in accord with the sound wave, $\tau = \frac{\hbar}{2\Lambda m \varrho_s}\left(\frac{\partial \mu_s}{\partial \varrho_s}\right)^{-1}_{p, s/\varrho}$ is the characteristic time of relaxation. The term $i\bar{\alpha}k^2$ describes the sound damping, $\bar{\beta}k^3$ — the dispersion.

On the free surface of superfluid helium, as on the surface of a common incompressible fluid, there can propagate the waves, the dispersion of which is affected by the forces of surface tension. The law of dispersion for such waves has the form [64]:

$$\omega^2 = (\varrho_s/\varrho)(\alpha/\varrho)k^3\,th(kH), \qquad \varrho = \varrho_s + \varrho_n$$

which coincides with the law of dispersion for ripple waves to the accuracy of the coefficient ϱ_s/ϱ.

The plasma, consisting of positive ions and negative electrons, is a dispersive medium for many types of wave processes [7, 62, 93].

Let us consider the ion-sound waves in a rarefied plasma with no magnetic field. Such a plasma is described by the following system of equations:

$$\frac{\partial n_{i,e}}{\partial t} + \text{div}\,(n_{i,e}\boldsymbol{u}_{i,e}) = 0,$$

$$\frac{\partial \boldsymbol{u}_{i,e}}{\partial t} + (\boldsymbol{u}_{i,e}\nabla)\,\boldsymbol{u}_{i,e} = -(n_{i,e}m_{i,e})^{-1}\,\nabla p_{i,e} - (q_{i,e}/m_{i,e})\,\nabla\varphi,$$

$$\varepsilon_0\,\Delta\varphi = -\,(q_i n_i + q_e n_e), \quad q_i = -q_e = e, \quad p_{i,e} = n_{i,e}T_{i,e}.$$

Here $n_{i,e}$ is the number of ions (electrons) in a unit volume, $u_{i,e}$ is the macroscopic velocity of ions (electrons), $p_{i,e}$ is the pressure of the ion (electron) gas, φ is the electrical potential. Let us consider the simplest case, when $T_e = \text{const.}$, $T_i = 0$. Then from the equation of motion for the electrons we can obtain the expression for the electron density: $n_e = n_0 \exp(e\varphi/T_e)$.

The ion gas is described by the equations

$$\frac{\partial \boldsymbol{u}_i}{\partial t} + (\boldsymbol{u}_i\nabla)\,\boldsymbol{u}_i = -(e/m_i)\,\nabla\varphi, \quad \frac{\partial n_i}{\partial t} + \text{div}\,(n_i\boldsymbol{u}_i) = 0. \quad (1.24)$$

These equations should be modified by the Poisson equation relating the potential to the ion density:

$$\varepsilon_0\,\Delta\varphi = -e\big(n_i - n_0 \exp(e\varphi/T_e)\big). \quad (1.25)$$

Having linearized equations (1.24), (1.25), we find the law of dispersion for plane ion-sound waves of small amplitudes:

$$\omega^2 = c_s^2 k^2 (1 + D^2 k^2)^{-1}, \quad v_{ph} = c_s(1 + D^2 k^2)^{-1/2} \quad (1.26)$$

where $c_s = (T_e/m_i)^{1/2}$, $D = (\varepsilon_0 T_e/n_0 e^2)^{1/2}$. Relations (1.26) demonstrate the phase velocity of small oscillations to decrease with decreasing wave length (with increasing wave number k) in an anisothermic plasma with no magnetic field, both for gravitational waves and the waves in a fluid with gas bubbles. Deviation of the dependence $\omega = \omega(k)$ from the linear one becomes essential at the wavelengths comparable with the Debye radius D. For the long waves $(kD \ll 1)$ we can easily get from (1.26):

$$\omega \approx c_s k \left(1 - \frac{1}{2} D^2 k^2\right),$$

$$v_{ph} \approx c_s \left(1 - \frac{1}{2} D^2 k^2\right), \tag{1.27}$$

$$v_g \approx c_s \left(1 - \frac{3}{2} D^2 k^2\right).$$

Now let us consider the magneto-sound waves in a 'cold' quasi-neutral rarefied plasma, immersed in a magnetic field. We refer to a plasma as 'cold' when its gas–kinetic pressure is considerably less than the magnetic pressure $(p \ll B^2/2\mu_0)$. The governing equations for such a plasma are as follows:

$$\frac{\partial n}{\partial t} + \text{div}\,(n u_{i,e}) = 0,$$

$$m_{i,e} \left(\frac{\partial u_{i,e}}{\partial t} + (u_{i,e} \nabla)\, u_{i,e}\right) = \pm e(E + [u_{i,e} B]), \tag{1.28}$$

$$\text{rot}\, B = \mu_0 e n(u_i - u_e), \quad \frac{\partial B}{\partial t} = -\text{rot}\, E, \quad \text{div}\, B = 0.$$

Let us direct the x-axis along the wave vector $\mathbf{k}$; let the undisturbed magnetic field B_0 lie in the plane x, z and be at the angle θ to the z-axis. Linearizing system (1.28), we get the following law of dispersion:

$$\omega = \frac{1}{2} V_A k (1 + \delta_e^2 k^2)^{-1/2} \{[(1 + \sin \theta)^2$$
$$+ \delta_i^2 k^2 (1 + \delta_e^2 k^2)^{-1} \sin^2 \theta]^{1/2} \pm [(1 - \sin \theta)^2 \tag{1.29}$$
$$+ \delta_i^2 k^2 (1 + \delta_e^2 k^2)^{-1} \sin^2 \theta]^{1/2}\}.$$

Here $V_A = B_0/(\mu_0 n_0 m_i)^{1/2}$ is the Alfven velocity, $\delta_e = c/\omega_{oe}$, $\delta_i = c/\omega_{oi}$, $\omega_{oe,i} = (n_0 e^2/\varepsilon_0 m_{e,i})^{1/2}$. The two branches, corresponding to the signs $\pm$ in (1.29), present fast and slow magneto-sound waves. Let us limit ourselves to the fast magneto-sound waves and write for them the law of dispersion in the limiting case of sufficiently long waves (setting, besides, that the angle $\theta \neq 90°$):

$$\omega \approx V_A k \left[1 - \frac{1}{2} \delta_i^2 (m_e/m_i - tg^2\theta) k^2 \right].$$

Therefore, at $0 \leq \theta < (m_e/m_i)^{1/2}$ the phase velocity of small oscillations reduces with growing wave number, which is analogous to the gravitational waves, those in a fluid with gas bubbles and ion-sound onesin a plasma with no magnetic field. At $(m_e/m_i)^{1/2} < \theta < 90°$, the phase velocity of small oscillations increases with growing wave number by analogy with the gravitation-ripple waves in 'shallow' water, when $H < (3\alpha/\varrho g)^{1/2}$.

If the equality $tg\theta = (m_e/m_i)^{1/2}$ (for the hydrogen plasma) holds, then we should take into account the next term of the expansion in the series of the wave number powers, since the coefficient at k^2 turns to zero and we get from (1.29)

$$\omega \approx V_A k \left(1 - \frac{1}{8} \delta_i^4 k^4 \right). \tag{1.30}$$

In the case when $\theta = 90°$, the two waves with circular polarization can propagate along the undisturbed magnetic field. In one of the waves, having the resonance at the frequency $\Omega_B = eB_0/m_i$, the vector of the electrical field E rotates in the direction of the ion rotation in the magnetic field, and in the low-frequency region ($\omega \lesssim \Omega_B$) for $\lambda \gg \delta_i$ we have

$$\omega \approx V_A k (1 - \delta_{ik})^{1/2}. \tag{1.31}$$

The phase velocity decreases with growing wave number, as was the case for the waves propagating across the magnetic field. In the other wave, having the resonance at the frequency $\omega_B = eB_0/m_e$, the vector E rotates in the direction of the electron rotation in the magnetic field, and in the low-frequency region ($\omega \lesssim \Omega_B$) for $\lambda \gg \delta_i$ we have

$$\omega \approx V_A k(1 + \delta_{ik})^{1/2}. \tag{1.32}$$

The phase velocity increases with growing wave number. In the high-frequency range ($\omega \gg \Omega_B$) the law of dispersion for these wave changes and the phase velocity of small oscillations decreases with growing wave number.

Note, that an ordinary gas, consisting of neutral molecules, has no dispersion with respect to acoustic oscillations in the absence of relaxation processes, chemical reactions and any thermodynamic nonequilibrium. For a neutral gas the speed of sound equals $c_s = (\gamma p/\varrho)^{1/2}$ and is independent from the wave number.

Some other examples of dispersive media can be found in the review [59]. In almost all of the examples cited, apart from the cases described by formulae (1.30)–(1.32), the law of dispersion for small-amplitude waves is the same (for the long waves as compared with certain characteristic spatial scales — asymptotically):

$$\omega = \bar{v}_{ph}k(1 \pm \delta^2 k^2), \tag{1.33}$$

where $\bar{v}_{ph} = \lim_{k\to 0} (\omega(k)/k)$ is the phase velocity of the infinitely long waves, δ is the so-called dispersion length, within which the dispersive effects reveal themselves. Knowing the law of dispersion, we can easily write the equation for any quantity in the wave. Indeed, in the linear approximation a partial solution is sought as $\exp i(kx - \omega t)$; then we set in the equations that $\dfrac{\partial}{\partial t} \to -i\omega, \dfrac{\partial}{\partial x} \to ik$, and, as a result, we get the relation $\omega = \omega(k)$. And vice versa, having the law of dispersion, we carry on the substitution $\omega \to i\dfrac{\partial}{\partial t}, k \to -i\dfrac{\partial}{\partial x}$, and, as a result, we get the equation

$$u_t + \bar{v}_{ph}u_x + \beta u_{xxx} = 0$$

where the coefficient $\beta = \pm\bar{v}_{ph}\delta^2$ can have a different sign. If we go over to the system of coordinates that moves with the velocity $\bar{v}_{ph}$, then we get

$$u_t + \beta u_{xxx} = 0. \tag{1.34}$$

The solution of this equation with the dispersion is expressed through the Airy's functions

$$u(x, t) = \pi^{-1/2}(3\beta t)^{-1/3} \int_{-\infty}^{\infty} Ai\left(\frac{x - x'}{(3\beta t)^{1/3}}\right) u(x', 0)\, dx',$$

$$Ai(z) = \pi^{-1/2} \int_{0}^{\infty} \cos\left(\frac{1}{3} y^2 + yz\right) dy$$

and is in a qualitative sense an oscillating wave packet. In the linear approach the wave packets, consisting of the waves of different wavelengths, expand in a dispersive medium in the course of time, since they propagate with different velocities.

If the law of dispersion is described by formula (1.30), instead of equation (1.34) we can get the equation

$$\frac{\partial u}{\partial t} + \bar{\beta}\, \frac{\partial^5 u}{\partial x^5} = 0. \tag{1.35}$$

In all the examples cited above, small disturbances of the medium parameters propagate without increasing the amplitude according either to the conventional wave equation $u_{tt} = c^2 u_{xx}$, or, in more complex cases to the equation with the dispersion $u_t + cu_x + \beta u_{xxx} = 0$. But there are a lot of examples when small disturbances increase in time with respect to the amplitude, which results in a sufficiently rapid violation of the validity of the linear approximation.

Let us consider the Alfven waves propagation in an anisotropic rarefied plasma. Such waves play a significant part in the process occurring in the solar wind plasma [63]. According to the data given in [63] the solar wind plasma is anisotropic: $p_\| \neq p_\perp$, where $p_\|$ and $p_\perp$ are the longitudinal and transversal pressures with respect to the undisturbed magnetic field B_0. A linear analysis of the kinetic equations given in [63] demonstrates that the Alfven waves propagating along the constant magnetic field B_0 in an anisotropic plasma with a high value of $\beta = 2\mu_0 p/B^2$ are unstable when the condition $p_\parallel > p_\perp + B_0^2/\mu_0$ is met. Since this criterion of instability of the Alfven waves does not depend on the detailed form of the function of particle distribution, but is determined by the macroscopic characteristics, it would be natural to expect that this instability, called the firehose instability [93, 63], can be sufficiently well described within the framework of the equations of hydrodynamic type, in which case, naturally, we limit ourselves to the waves with $\lambda \gg R$ (where R is the Larmor radius of the particles). Such an instability reveals itself in the case when the centrifugal force, affecting the particles in their motion along the curved line of force, exceeds the restoring force, associated with the tension of the lines of force.

As a hydrodynamic model of a rarefied anisotropic plasma, use is often made of the model by Chew–Goldberger–Low [40] (hereafter referred to as the CGL model) and of its generalizations, accounting for the finite value of the Larmor ion radius [89, 112].

Let us consider the propagation of the transversal Alfven waves along the undisturbed magnetic field $\boldsymbol{B}_0$, the direction of which coincides with the z-axis [27]. Let us assume that all the functions depend only on the coordinate z and the time t. In this case the corresponding equations are as follows:

$$\varrho \frac{\partial u}{\partial t} + \frac{\partial}{\partial z} \left\{ (p_\parallel - p_\perp) \frac{B_0 B_x}{B^2} - B_0 B_x/\mu_0 - \Omega_B^{-1} \right.$$

$$\left. \left(p_\perp + \frac{p_\parallel - p_\perp}{B^2} B_0^2 \right) \frac{\partial v}{\partial z} \right\} = 0,$$

$$\varrho \frac{\partial v}{\partial t} + \frac{\partial}{\partial z} \left\{ (p_\parallel - p_\perp) \frac{B_0 B_y}{B^2} - B_0 B_y/\mu_0 \right.$$

$$\left. + \Omega_B^{-1} \left(p_\perp + \frac{p_\parallel - p_\perp}{B^2} B_0^2 \right) \frac{\partial u}{\partial z} \right\} = 0,$$

$$\frac{\partial B_x}{\partial t} = B_0 \frac{\partial u}{\partial z}, \qquad \frac{\partial B_y}{\partial t} = B_0 \frac{\partial v}{\partial z}, \qquad (1.36)$$

$$\frac{\partial}{\partial t}(p_\parallel B^2) = 0, \qquad \frac{\partial}{\partial t}(p_\perp/B) = 0.$$

Here u, v are the x-, y-components of the plasma velocity, $\varrho = \text{const.}$ is the plasma density, $B^2 = B_x^2 + B_y^2 + B_0^2$, $\Omega_B = eB_0/m_i$ is the ion cyclotron frequency. The terms that are proportional to the quantity Ω_B^{-1} describe the so-called magnetic (non-dissipative) viscosity.

If the problem is considered in the interval $0 \div L$ with the periodical boundary conditions, then the law of the energy conservation for the processes described by system (1.36) can be written in the following way:

$$\int_0^L \left(\frac{1}{2} \varrho(u^2 + v^2) + \frac{1}{2} p_\parallel + p_\perp + B^2/2\mu_0 \right) dz = \int_0^L Q(z, t) \, dz = \text{const.}$$

The total energy density $Q(z, t)$ consists of three parts: the kinetic energy density $\frac{1}{2} \varrho(u^2 + v^2)$, the internal energy density $\frac{1}{2} p_\parallel + p_\perp$ and the magnetic energy density $B^2/2\mu_0$. In an isotropic gas the internal energy density is determined by the formula $\varepsilon = p/(\gamma - 1)$, $\gamma = (m + 2)/m$, where m is the number of the degrees of freedom of the gas molecules. In the considered case of an anisotropic plasma, the particle motion along the magnetic field can be viewed as one-dimensional when $m = 1$, $\gamma = 3$, and the particle motion across the magnetic field can be regarded as two-dimensional when $m = 2$, $\gamma = 2$. Therefore, the density of the internal energy associated with the longitudinal pressure is equal to $p_\parallel/2$, while that associated with the transversal pressure equals $p_\perp$.

Linear analysis of the system of equations (1.36) readily affords the following dispersive equation:

$$\omega_k = \omega^{(k)} + i\gamma_k,$$

$$\omega^{(k)} = \frac{1}{2} \Omega_B(kR)^2, \qquad (1.37)$$

$$\gamma_k = \Omega_B kR \left[p_\parallel^{-1}(p_\parallel - p_\perp - B_0^2/\mu_0) - \frac{1}{4}(kR)^2 \right]^{1/2},$$

where $R = \Omega_B^{-1}(p_\parallel/\varrho)^{1/2}$ is the Larmor ion radius, k is the wave number of small disturbances, γ_k is their growth rate. Expression (1.37) for the increment demonstrates the considered Alfven waves of small amplitudes to be unstable under the following conditions:

$$p_\| > p_\perp + B_0^2/\mu_0, \qquad (1.38)$$

$$kR < 2[p_\|^{-1}(p_\| - p_\perp - B_0^2/\mu_0)]^{1/2}. \qquad (1.39)$$

Condition (1.38) shows that the firehose stability arises at a noticeable plasma anisotropy. Within the CGL model and with the Larmor ion radius accounted for, condition (1.39) demonstrates that sufficiently long Alfven waves are unstable, while the short-wave harmonics with $\lambda < \pi R\left[p_\|\left(p_\| - p_\perp - \dfrac{B_0^2}{\mu_0}\right)^{-1}\right]^{1/2}$ get stabilized. The increment γ_k is maximum for the harmonics with $kR = [2p_\|^{-1}(p_\| - p_\perp - B_0^2/\mu_0)]^{1/2}$, and becomes zero at $kR = 2[p_\|^{-1}(p_\| - p_\perp - B_0^2/\mu_0)]^{1/2}$. Therefore, in an anisotropic plasma there arise the unstable Alfven waves for which $0 \lesssim kR \lesssim 2\left[p_\|^{-1}\left(p_\| - p_\perp - \dfrac{B_0^2}{\mu_0}\right)\right]^{1/2}$. Since the hydrodynamic approach is valid at $kR \ll 1$, the firehose instability is of a hydrodynamic character at small degrees of the plasma anisotropy $\Delta p \equiv p_\| - p_\perp - B_0^2/\mu_0 \ll p_\|$.

Quite a similar picture of instability within a limited spectrum region (for the wavelengths greater than a certain value of λ_*) arises when studying viscous films falling along solid oblique planes. Let us direct the x-axis along a solid surface, the y-axis — along the normal to it, and let $u = \{u, v, w\}$ be the fluid velocity, p — the pressure, $\varrho = \text{const.}$ — the density (the fluid is considered incompressible). In the case in question, the Navier–Stokes equations have the steady solution

$$\bar{u} = (g/\nu)\left(h_0 y - \frac{1}{2}y^2\right)\sin\theta, \quad \bar{v} = \bar{w} = 0,$$

$$\bar{p} = \pi - \varrho g(y - h_0)\cos\theta, \qquad (1.40)$$

where h_0 is the film thickness, ν is the viscosity coefficient, π is the atmospheric pressure, g is the gravitational acceleration, and $\theta \neq 0$ is the angle of the plane inclination to the horizon. Let us introduce the small disturbances $u = \bar{u} + u'$, $v = \bar{v} + v'$, $w = \bar{w} + w'$, linearize the Navier–Stokes equations and make use of the condition of the continuity of the normal pressure on a free surface, of the tangential stress equality to zero, of the no-slip condition on a solid wall, as well as of the kinematic condition on a free surface. As a result, we get the following linear equation in dimensionless variables:

$$\zeta_t + 3\zeta_x + \alpha\zeta_{xx} + \beta\zeta_{xxxx} = 0, \tag{1.41}$$

where $\zeta(x, t) = h(x, t) - h_0$ is an elevation of the film surface over the undisturbed level, $Re = U_0 h_0 v^{-1}$ is the Reynolds number, $U_0 = (gh_0^2/3v) \sin \theta$, $\mathscr{W} = T/\varrho gh_0^2$ is the Weber number (T is the surface tension coefficient), $\alpha = \dfrac{6}{5} Re - ctg\theta$, $\beta = \mathscr{W}/\sin \theta$. In the system of the coordinates moving with the velocity 3, the law of dispersion for plane waves $\exp i(kx - \omega_k t)$ will have the following form:

$$\omega_k = \omega^{(k)} + i\gamma_k, \quad \omega^{(k)} = 0, \quad \gamma_k = k^2(\alpha - \beta k^2). \tag{1.42}$$

The expression for the growth rate demonstrates that the film surface is unstable under the two following conditions:

$$Re > \frac{5}{6} ctg \, \theta,$$

$$k < (\alpha/\beta)^{1/2} = \left(\frac{6Re - 5ctg\theta}{5\mathscr{W}} \sin \theta\right)^{1/2}, \tag{1.43}$$

i.e. as was the case with the firehose instability, the short-wave disturbances with $\lambda < 2\pi(\beta/\alpha)^{1/2}$ get stabilized. The harmonics with the wave number $k = (\alpha/2\beta)^{1/2}$ and with its increment equal to $\gamma_{max} = \alpha^2/4\beta$ is the fastest growing one. At $\theta \neq 90°$ this instability is a threshold one with respect to the Reynolds number, while at $\theta = 90°$ it occurs at any Reynolds numbers.

Chapter 2

Finite-Amplitude Waves in Weak Dispersive and Unstable Media

2.1. One-Dimensional Solitons

A simplest model equation for analysing the evolution of the waves of a small but finite amplitude in a dispersive medium can be obtained by adding to the linear equation (1.34) the term uu_x, describing a convective transfer:

$$u_t + uu_x + \beta u_{xxx} = 0. \qquad (2.1)$$

This equation was obtained by Korteweg and de Vries as far back as 1895 by way of expanding the equations of an ideal incompressible fluid in the small parameters $a/\lambda \ll 1$, $H/\lambda \ll 1$, where a is the wave amplitude, H is the channel depth, λ is the wavelength. But it was only from the beginning of the 1960s, after R. Z. Sagdeev discovered and formulated the determining role of dispersive effects in forming non-linear waves in dispersive media [93], that due attention has been paid to this equation.

Equation (2.1) that is now called the Korteweg–de Vries equation (KdV), has been obtained for the magneto-sound waves propagating either perpendicular to the magnetic field or at angles close to 90° in a cold rarefied plasma [20]. The equation was derived based on the expansion of the governing equations, taking no account of all the dissipative processes and the pressure to the accuracy of the quadratic terms with respect to the amplitude of small disturbances, and on the approximate factorization of the wave operator, which made it possible to reduce the order of the equation. In a number of works (see, for instance, [79, 108]), there has been developed a formalized approach for deducing model equations of type (2.1) from the governing equations, describing dispersive media, by way of transforming the arguments x, t and expanding in the small parameter; the method of quasi-simple waves [62] being the most typical of this kind. The basic suppositions made when deducing equation (2.1) are the following: (1) a small but finite amplitude; (2) the wavelength is great as compared with the dispersion length (see Chapter 1). To date the existence and uniqueness of the solution of the KdV equation has been proved [59].

The laws of dispersion demonstrated earlier show that in a general case the dispersion parameter β can be both positive (when the phase velocity v_{ph} decreases with growing wave number) and negative (when v_{ph} increases with growing wave number). However, it can be easily demonstrated that the solutions of the KdV equation with $\beta < 0$ are obtained from those with $\beta > 0$ by substituting $x \to -x$, $u \to -u$, and, hence, hereafter the dispersion parameter β will be considered positive.

If the dispersion is absent or can be neglected, then $\beta = 0$ and the equation $u_t + uu_x = 0$ is known to describe the evolution of the initial disturbances, having the areas where $\partial u / \partial x < 0$, with a continuous increase of the steepness of the leading edges of the profile in this areas.

Indeed, let us consider the case of small but finite amplitudes and express the required function as a series with respect to a small parameter: $u(x, t) = \mu u_1(x, t) + \mu^2 u_2(x, t) + \ldots$ [60]. Substituting the resulting series into the equation with the terms at the same degrees equated, we get the following results: $\partial u_1 / \partial t = 0$, $u_1(x, t) = A \sin kx$ — in the first approximation the initial, for instance, sinusoidal wave propagates without changing its form;

$$\frac{\partial u_2}{\partial t} = -\frac{1}{2} k A^2 \sin 2kx, \quad u_2(x, t) = -\frac{1}{2} k A^2 t \sin 2kx;$$

taking into account only these terms, we have:

$$u(x, t) = A \sin kx - \frac{1}{2} kA^2 t \sin 2kx. \tag{2.2}$$

Account taken of the non-linearity results in the appearance of the function $u_2(x, t)$ and reflects a distortion of the initial sinusoidal wave $u_1(x, t) = A \sin kx$: there arises the second harmonic $A' \sin 2kx$ with half the wavelength, with the amplitude $A' = \frac{1}{2} kA^2 t$ increasing linearly with growing time; there occurs an increase in the steepness of the initial profile in the range where $\partial u/\partial x < 0$ to the extent of forming a discontinuity with $|\partial u/\partial x| = \infty$. Formula (2.2) is valid only at sufficiently small times, since at $t \sim (kA)^{-1}$ the function u_2 becomes comparable with the function u_1 with respect to the order of magnitude, and the method of iterations is of no use. The situation with $|\partial u/\partial x| = \infty$, that has no physical sense, is eliminated by introducing the viscosity (dissipation) or dispersion.

Dissipation results in a model Burgers equation $u_t + uu_x = \alpha u_{xx}$ ($\alpha > 0$ is the viscosity coefficient), which has a well-known stationary solution — a 'shock wave' with the width of its forefront $\delta \sim \alpha/U$, where $U = \frac{1}{2}(u_{-\infty} + u_{\infty})$, since the non-linearity effect is compensated for by the smearing effect of viscosity.

Dispersion can be also compensated for by the effect of non-linearity. In line with [60], let us limit ourselves, as before, by the first two harmonics: u_1 and u_2, the phase velocities of which are different in the presence of dispersion, i.e. $v_{ph}^{(1)} \neq v_{ph}^{(2)}$. In the system of coordinates, moving with the velocity $v_{ph}^{(1)}$, the equation for the second harmonic is as follows:

$$\frac{\partial u_2}{\partial t} + \delta v_{ph} \frac{\partial u_2}{\partial x} = -\frac{1}{2} kA^2 \sin 2kx, \qquad \delta v_{ph} = v_{ph}^{(2)} - v_{ph}^{(1)}.$$

Setting $u_2(x, 0) = 0$, we get the solution of this equation:

$$u_2(x, t) = -(A^2/2\delta v_{ph}) \sin (\delta v_{ph} t) \sin (2kx - \delta v_{ph} t)$$

which demonstrates that in the course of time the solution does not increase, and the amplitude of the second harmonic remains a quantity

of the second order of smallness with respect to the amplitude of the
first harmonic A.

Counter-balancing the non-linearity effect, the dispersion makes it
possible for the steady waves of a finite amplitude — solitary and
periodic — to be formed in a dispersive medium. Korteweg and de
Vries established [65] that solitary waves, that have been referred to
as solitons since the publication of [111], are described by the formula

$$u(x, t) = u_0 ch^{-2}(\xi/l_s),$$

$$\xi = x - Ut, \qquad U = u_0/3, \qquad l_s = (12\beta/u_0)^{1/2}. \qquad (2.3)$$

A soliton propagates without changing its form with a constant velocity $U = u_0/3$, its width l_s decreases with its growing amplitude u_0,
solitons with greater amplitudes move faster. Formula (2.3) is ob-

tained by replacing in (2.1) $\dfrac{\partial}{\partial x} \rightarrow -U\dfrac{d}{d\xi}, \dfrac{\partial}{\partial x} \rightarrow \dfrac{d}{d\xi}$ and by integrating

an ordinary differential equation of the third order, with allowances
made for the fact that the sought function and the necessary number
of its derivatives become zero in the infinity. Periodic waves are called
the cnoidal ones, the corresponding formulae given in [65].

The first unsteady (similarity) solution of the equation was obtained
and studied in [19, 20]:

$$u(x, t) = \beta(3\beta t)^{-2/3} \, \psi\{(3\beta t)^{-1/3} \, x\}. \qquad (2.4)$$

The function ψ of the similarity variable $z = (3\beta t)^{-1/3} \, x$ obeys the
equation of the third order (the prime denotes differentiation with
respect to z):

$$\psi''' + \psi\psi' - z\psi' - z\psi = 0 \qquad (2.5)$$

which through the use of the substitution $\psi = \varphi' - \dfrac{1}{6}\varphi^2$ is reduced
to the equation of the second order

$$\varphi'' = z\varphi + \frac{1}{18}\varphi^3 \qquad (2.6)$$

for the new function $\varphi(z)$ related to the function $\psi(z)$ through the equation $\varphi' - \dfrac{1}{6}\varphi^2 = \psi$.

The roots of the characteristic equation, corresponding to (2.6) are real and of different signs at $z > 0$, and imaginary (conjugated) at $z < 0$. Hence, the singular points of the equation (2.6) are the saddle at the positive values of the argument z and the centre at its negative values. The linearized equation $\varphi'' = z\varphi$ is an Airy's equation and at all z has a finite solution with the asymptotics

$$\varphi(z) = \frac{1}{2} C z^{-1/4} \exp\left(-\frac{2}{3} z^{3/2}\right) \quad \text{at} \quad z \to \infty,$$

$$\varphi(z) = C |z|^{-1/4} \sin\left(\frac{2}{3} |z|^{3/2} + \frac{\pi}{4}\right) \quad \text{at} \quad z \to -\infty,$$

where C is an arbitrary constant.

The asymptotics of the function $\psi(z)$ has the form

$$\psi(z) = -\frac{1}{2} C z^{1/4} \exp\left(-\frac{2}{3} z^{3/2}\right) \quad \text{at} \quad z \to \infty,$$

$$\psi(z) = -C |z|^{1/4} \cos\left(\frac{2}{3} |z|^{3/2} + \frac{\pi}{4}\right) \quad \text{at} \quad z \to -\infty. \tag{2.7}$$

If in formula (2.7) the constant C is chosen sufficiently small, then the solution of the non-linear equation (2.5) at $z \to -\infty$ can be sought in the form

$$\psi(z) = C_1(z) \cos\left(\frac{2}{3} |z|^{3/2} + \frac{\pi}{4}\right) + C_2(z) \sin\left(\frac{2}{3} |z|^{3/2} + \frac{\pi}{4}\right),$$

where $C_1(z)$, $C_2(z)$ are the slowly changing functions of z. Substituting this form of the solution into equation (2.5), at $z \to -\infty$ we get

$$\psi(z) = |z|^{1/4} \left\{ C_1 \cos\left(\frac{2}{3} |z|^{3/2} + \frac{\pi}{4}\right) + C_2 \sin\left(\frac{2}{3} |z|^{3/2} + \frac{\pi}{4}\right) \right\} + C_3 |z|^{-1/2},$$

where C_1, C_2, C_3 are the constants; the solution oscillates with a slowly growing amplitude. In [20] the solution was shown to be of this type at $C < C_*$ (in the variables used in the present monograph, $C_* \approx 3.35$). At $C > C_*$ the solution of the non-linear equation (2.5) has a pole of the form $(z + \delta)^{-2}$.

Unsteady solutions of the KdV equation, describing the evolution of the initial disturbances of various types (a sine curve, a Gaussian curve, etc.) were first numerically obtained in [8, 18, 111]. It is these numerical investigations that made it possible to get the first (and basic) qualitative and quantitative data on the character of unsteady flows described by KdV, and provided useful insight into understanding these phenomena and choosing the analytical approach developed later [51].

Let us consider the evolution of the localized initial disturbances ($|u(x, 0)| \rightarrow 0$ at $|x| \rightarrow \infty$) in a dispersive medium on the basis of the KdV equation. Let $u(x, 0) = u_0 f(x)$. In equation (2.1) let us go over to dimensionless parameters, choosing the value u_0 as a characteristic velocity, and the 'width' of the initial disturbance l as a characteristic space scale. As a result, we get the following dimensionless equation:

$$u_t + uu_x + \sigma^{-2}u_{xxx} = 0, \quad u(x, 0) = f(x). \tag{2.8}$$

Here $\sigma = (u_0 l^2/\beta)^{1/2}$ is the similarity parameter, playing in dispersive media the same role as the Reynolds number in dissipative (viscous) media, and characterizing the degree of the problem, non-linearity. Substituting the soliton 'width' l_s into the similarity parameter, gives $\sigma = \sigma_s = 2\sqrt{3}$. This value of the parameter is a characteristic one: at $\sigma \gg \sigma_s$ the initial disturbance with $u(x, 0) > 0$ breaks down practically only into solitons (the greater the value of σ, the more solitons are produced); at $\sigma \leqslant \sigma_s$ the onlysoliton, accompanied, with growing time, by expansion of a wave packet of a considerably less amplitude, is formed in the front portion of the profile.

If the initial disturbance breaks down only into solitons, the conditions of the disturbance decompose into a definite number of solitons and their amplitudes can be approximately determined by the laws of conservation [18]:

$$\int\limits_{-\infty}^{\infty} u(x, t)\, dx, \qquad \int\limits_{-\infty}^{\infty} \frac{1}{2} u^2(x, t)\, dx, \qquad \int\limits_{-\infty}^{\infty} \left(\frac{1}{3} u^3 - \beta u_x^2 \right) dx, \ldots .$$

The KdV equation has an infinite number of the laws of conservation

$$\frac{\partial Q_m(x, t)}{\partial t} + \frac{\partial P_m(x, t)}{\partial x} = 0, \qquad m = 1, 2, \ldots . \qquad (2.9)$$

For the first three laws we get

$$Q_1 = u, \qquad P_1 = \frac{1}{2} u^2 + \beta u_{xx},$$

$$Q_2 = \frac{1}{2} u^2, \qquad P_2 = \frac{1}{3} u^3 + \beta \left(u u_x - \frac{1}{2} u_x^2 \right),$$

$$Q_3 = \frac{1}{3} u^3 - \beta u_x^2, \qquad P_3 = \frac{1}{4} u^4$$

$$+ \beta(u^2 u_{xx} + 2 u_t u_x) + \beta^2 u_{xx}^2 .$$

Integrating equation (2.9) with respect to the spatial coordinate and assuming damping of the disturbances at $x \to \pm\infty$, we get

$$\int\limits_{-\infty}^{\infty} Q_m(x, t)\, dx = \int\limits_{-\infty}^{\infty} Q_m(x, 0)\, dx = S_m .$$

If the initial disturbance produces N solitons, then, after they have gone sufficiently far in separate ways, the invariants S_m will be the sum $\sum\limits_{n=1}^{N} S_m^{(n)}$ of the invariants of the individual solitons, and the following equation can be written

$$\sum_{n=1}^{N} A_n^{m-1/2} = (\sigma/\sigma_s) \int\limits_{-\infty}^{\infty} Q_m(x, 0)\, dx \Big/ \int\limits_{-\infty}^{\infty} q_m(x)\, dx, \qquad m = 1, \cdots, N$$

$$(2.10)$$

where A_n is the soliton dimensionless amplitude, $Q_m(x, 0)$ and $q_m(x)$ are the invariant densities for the initial function and the soliton. The system of equations (2.10) allows one to determine the soliton amplitudes. Let the initial disturbance produce two solitons. In this case, solving system (2.10) for $N = 2$, $m = 1, 2$, we get the following expressions for the amplitudes:

$$A_{1,2} = (1/4) [a_1 \pm (4a_2/3a_1 - a_1^2/3)^{1/2}]^2$$

where

$$a_1 = (\sigma/2\sigma_s) \int\limits_{-\infty}^{\infty} f(x)\, \mathrm{d}x, \quad a_2 = (3\sigma/4\sigma_s) \int\limits_{-\infty}^{\infty} f^2(x)\, \mathrm{d}x.$$

From the condition of reality and positiveness of the soliton amplitudes it follows that for the initial disturbance with $S_1 > 0$ to break down into two solitons, the condition $\sigma_c < \sigma < 2\sigma_c$ must be met, wherein

$$\sigma_c^2 = 6\sigma_s^2 \int\limits_{-\infty}^{\infty} f^2(x)\, \mathrm{d}x \Big/ \left(\int\limits_{-\infty}^{\infty} f(x)\, \mathrm{d}x \right)^3.$$

If $f(x) = \exp(-x^2)$, then $\sigma_c = (6\sqrt{2}\,\pi)^{1/2} \sigma_s \approx 4$, if $f(x) = ch^{-2}x$, then $\sigma_c = \sigma_s$. The value of the similarity parameter $\sigma = \sigma_c$ indicates the difference between the disturbances breaking down into solitons and those producing only one soliton and an oscillating wave packet. At $\sigma > \sigma_c$ the initial compression $f(x) = \exp(-x^2)$ always decomposes into solitons: at $\sigma \approx 4$ two solitons are produced, at $\sigma \approx 8$ — three solitons, at $\sigma \approx 12$ — four solitons and so on, as has been discovered and confirmed by numerical calculations [18].

A numerical solution of the KdV with the initial condition $f(x) = \exp(-x^2)$ within the range of changes in the similarity parameter $0.5 \le \sigma < 4$ demonstrates that at $\sigma < \sigma_c$ the character of the evolution does not depend on the parameter σ value in a qualitative respect. In line with the law of dispersion, at small times the oscillations with a negative velocity are formed (it should be recalled that the KdV equation is written in the system of coordinates moving with respect to the medium with the velocity $\bar{v}_{ph} = \lim\limits_{k \to 0} (\omega(k)/k) = \max\limits_{k} (\omega(k)/k)$), their number increasing and the wave packet expanding with growing time. In the process of evolution the front portion of the disturbance profile produces a soliton, moving towards positive values of the coordinate x with the velocity $u_{max}/3$ without changing its shape. Analysis of the results of the KdV numerical solutions has revealed a very interesting fact: the produced soliton carries away a greater portion of the momentum and the disturbance energy.

In a one-dimensional case, solitons are absolutely stable formations, preserving their shape in the conditions of non-linear interactions. To verify this fact, certain calculations have been carried out [8] with $f(x) = ch^{-2}(x/l)$, where $l < l_s$, $\sigma < \sigma_s$; at first the dispersive effects predominate over the non-linear ones, which results in the formation of a going-back wave packet, a decrease in the amplitude and a certain increase in the width of the front portion of the profile. As soon as this width reaches the soliton value l_s a decrease in the amplitude ceases and a soliton is formed. Soliton stability has also been confirmed by way of calculating the interaction of two solitons of different amplitudes.

If, however, the amplitude of the soliton moving along the x-axis is a function of the coordinate y (there exists an amplitude modulation), the problem becomes two-dimensional, and, depending on the law of dispersion (the sign of the parameter β) the soliton can be either stable (at $\beta > 0$) or unstable (at $\beta < 0$) [61].

If the initial disturbance is such that its first invariant is $S_1 \leq 0$, then the decomposition of this disturbance only into solitons is impossible, since the area of the profile, or the first invariant, is preserved, while for a soliton it must be positive (it should be recalled that only the case of $\beta > 0$ is considered throughout).

A numerical solution of the KdV with the initial condition

$$u(x, 0) = (\beta\sigma^2/\sqrt{\pi}) \frac{\partial}{\partial x} (e^{-x^2/l^2}),$$ for which $S_1 = 0$, demonstrates that

such disturbances evolutionize in an essentially different way, depending on the similarity parameter σ: at $\sigma < \sigma_0 \approx 3$ a dispersive spreading of the disturbance takes place, no solitons are formed, which results in a quasi-linear solution (by the terminology of [62]); at $\sigma > \sigma_0$ the solution is of a mixed type (an intensive wave packet and the solitons whose number increases with growing σ). Estimation of the critical similarity parameter σ_0 which serves a borderline between the quasi-linear solutions, resembling the similarity solution (2.4), and the solutions of a mixed type, has resulted in $\sigma_0 \approx 2.5$ [7]. This value is in accord with the data presented in [62], where by means of an approximate analytical method it has been demonstrated that the evolution character of the disturbances with the zero first invariant is different at $\sigma < \sigma_0$ and $\sigma > \sigma_0 \approx 3$: in the former case the solutions are quasi-linear while in the latter case solitons arise. The results discussed above refer only to the initial disturbances localized in a limited domain, i.e. $|u(x\ 0)| \to 0$ at $|x| \to \infty$.

In [10, 106] one can find the numerical solutions of the KdV equation with the initial condition $u(x\ 0) = C[1 + \exp{(x/l)}]^{-1}]$ where the quantity l sets the width of the forefront of the initial disturbance. The limiting transition $l \to 0$ makes it possible to obtain a set of disturbances tending to a discontinuity: $u(x, 0) = C$ at $x < 0$ and $u(x, 0) = 0$ at $x > 0$. The evolution of such a ramp is accompanied by a constant formation of new oscillations. The amplitude of the front oscillation reaches the value $2C$ within a finite time that is inversely proportional to the width of the initial step, while its velocity equals a third of the amplitude to a sufficient accuracy (which is analogous to the case of the solitons produced by the initial limited disturbances). According to numerical calculations, as soon as the amplitude of the front soliton gets equal to $2C$, the amplitude of the subsequent soliton also becomes equal to $2C$, and so on. Therefore, with growing time there occurs a successive levelling of the front solitons amplitudes up to the value $2C$, and one can assume that at sufficiently great x, t a train of solitons of the same amplitude, moving with a constant velocity $U = \dfrac{2}{3} C$, is formed.

If a dispersive medium is characterized by a certain energy dissipation (for instance, viscosity, heat conductivity, a finite electric conductivity), then the simplest model to describe it is the Burgers–Korteweg–de Vries equation (BKV):

$$u_t + u u_x + \beta u_{xxx} = \alpha u_{xx}, \quad \alpha > 0, \quad \beta > 0. \quad (2.11)$$

At $\beta = 0$, the equation (2.11) has a steady solution, which is a shock wave with the width of the forefront $\Delta \sim \alpha/U$, $U = \dfrac{1}{2}(u_{-\infty} + u_{\infty})$.

The BKV equation also has a steady solution in the form of a shock wave, but in this case the existence of two different structures, i.e. oscillatory and monotonic, is possible. Trying to find the steady solution $u(x, t) = u(x - Ut)$ let us denote $u_{-\infty} = u_1$, $u_\infty = u_0$, $u_1 > u_0$. in which case the wave velocity $U = \dfrac{1}{2}(u_1 + u_0)$. By way of linearizing the equation (2.11) in the vicinity of the singular points $u = u_0$, $u = u_1$, we shall find the following expressions for the roots of the characteric equation:

$$k_{0,1} = (\alpha/2\beta)\,[1 \pm (1 - 4\alpha^{-2}\beta(u_{0,1} - U))^{1/2}].$$

In the singular point '0' (at $u = u_0$)

$$k_0 = (\alpha/2\beta)\,[1 \pm (1 + 2\alpha^{-2}\beta(u_1 - u_0))^{1/2}]$$

which means that the roots in the singular point are real and of different signs; therefore, the singular point '0', corresponding to the undisturbed state of the gas in front of the wave, is always a saddle point and the integral curve goes off the point '0' monotonically.

In the singular point '1', corresponding to the disturbed state of the gas after the wave, we have

$$k_1 = (\alpha/2\beta)\,(1 \pm (1 - 2\alpha^{-2}\beta(u_1 - u_0))^{1/2}]. \qquad (2.12)$$

Provided the wave amplitude $u_1 - u_0$ is set, the above formula shows that: (1) if $2\alpha^{-2}\beta(u_1 - u_0) > 1$, then the roots k_1 are complex-conjugate, the singular point '1' is a focus, and the integral curve approaches this point with oscillations; (2) if $2\alpha^{-2}\beta(u_1 - u_0) < 1$, then the roots k_1 are real and of the same sign, the singular point '1' is a node, and the integral curve approaches this point monotonically. Thus, at the set wave amplitude $u_1 - u_0$ there exists such a critical value of the viscosity coefficient $\alpha = \alpha_* = [2\beta(u_1 - u_0)]^{1/2}$ that at $\alpha < \alpha_*$ (small viscosity) there is a shock wave with the oscillatory structure, while at $\alpha > \alpha_*$ (great viscosity) it is a shock wave with a monotonous structure.

Computation results of the unsteady BKV equation have revealed the fact that the formation of solitons, oscillating wave packets and 'viscous' triangle profiles depends on the relation σ^2/Re ($Re \sim \alpha^{-1}$ is the Reynolds number) [7, 10, 52]. The results given in this paragraph are general for all dispersive media and wave processes, when the law of dispersion has the form (1.33) and the amplitudes are finite but small. Experimental results on wave propagation in gas–fluid mixtures are in good agreement with the conclusions resulting from solving the KdV and BKV equations [69, 70].

By way of concluding the consideration of the KdV and BKV models, let us note that the character of the law of dispersion allows one to make conclusions on qualitative peculiarities of the non-linear waves of not a small amplitude as well, when the limits of these equations applicability are violated, i.e. to state whether the non-linear wave is a compression ($\beta > 0$) or a rarefaction ($\beta < 0$), whether the non-linear wave has an oscillatory 'tail', falling behind the basic profile ($\beta > 0$), or an oscillatory 'precursor', going in front of the basic profile ($\beta < 0$), to determine a spatial scale of the non-linear wave, which is equal to the dispersion length by the order of magnitude. It stands to reason that in a general case of more complex wave processes and, accordingly, the laws of dispersion, it is not the case; besides, the laws of dispersion, naturally, cannot indicate the phenomenon of overturning of non-linear waves that takes place at critical amplitudes [93].

The notion of dispersion, introduced into mechanics and physics, the KdV and BKV model equations and the analysis of their solution proved useful when investigating finite-difference algorithms and interpreting numerical solutions of the gas-dynamic and hydrodynamic problems, obtained through them. We shall limit our consideration to some simple difference schemes for the linear and quasi-linear transfer equations $u_t + u_x = 0$, $u_t + uu_x = 0$ with the initial condition $u(x, 0) = 1$ at $x \leq 0$ and $u(x, 0) = 0$ at $x > 0$.

$$(1) \quad u_t + u_x = 0, \quad u(x, 0) = u_0(x) = \begin{cases} 1 \text{ at } x \leq 0, \\ 0 \text{ at } x > 0. \end{cases}$$

Let us write for this equation the explicit scheme

$$u_j^{n+1} = u_j^n - (\tau/h)\,(u_j^n - u_{j-1}^n),$$

the first differential approximation of which has the form:

$$u_t + u_x = \frac{1}{2}\,(hu_{xx} - \tau u_{tt}).$$

In the solution of the governing equation $u_{tt} = u_{xx}$, hence, $u_t + u_x = \frac{1}{2}\,(h - \tau)\,u_{xx}$, or, assuming $\tau = kh, k = \text{const.}$, $u_t + u_x = \frac{h}{2}\,(1 - k)u_{xx}$. The order of approximation is $0(\tau, h)$; for the stability $\tau < h$, therefore, the coefficient of 'numerical viscosity', equal to $\frac{1}{2}\,(1 - k)\,h$, is positive. Thus, the explicit difference scheme has a numerical viscosity, which results in smoothing the solutions with strong gradients.

Let us consider the implicit scheme

$$u_j^{n+1} = \left(1 + \frac{\tau}{h}\right)^{-1}\left(u_j^n + \frac{\tau}{h}\,u_{j-1}^{n+1}\right).$$

This scheme, being formally implicit, is realized as an explicit one, since for the numerical solution it is necessary to set a boundary condition on the left boundary of the interval, whereon the problem is being solved. The first differential approximation is as follows

$$u_t + u_x = \frac{1}{2}\,(h + \tau)\,u_{xx} = \frac{1}{2}\,(1 + k)\,hu_{xx}.$$

The coefficient of numerical viscosity $\frac{1}{2}\,(1 + k)\,h$ is greater than that for the explicit scheme, which results in greater smoothing of the solutions.

For the Lax–Wendroff scheme, where

$$u_{j+1/2}^{n+1/2} = \frac{1}{2}\,(u_j^n + u_{j+1}^n) - \frac{\tau}{2h}\,(u_{j+1}^n - u_j^n) \quad \text{is a predictor,}$$

$$u_j^{n+1} = u_j^n - \frac{\tau}{h}\,(u_{j+1/2}^{n+1/2} - u_{j-1/2}^{n+1/2}) \quad \text{is a corrector,}$$

by way of excluding the fractional step, we get

$$u_j^{n+1} = u_j^n - \frac{\tau}{2h}(u_{j+1}^n - u_{j-1}^n) + \frac{\tau}{2h^2}(u_{j+1}^n - 2u_j^n + u_{j-1}^n).$$

The first differential approximation is as follows

$$u_t + u_x + \frac{1}{6}(h^2 - \tau^2)u_{xxx} = 0.$$

This linear equation is a dispersive one, therefore, the Lax–Wendroff scheme possesses a 'numerical dispersion', which results in the appearance of dispersive oscillations in the difference solution. Unlike the two former schemes with the order of approximation $0(\tau, h)$ the Lax-Wendroff scheme, in spite of the fact that its order of approximation is $0(\tau^2, h^2)$, introduces some non-monotonicities absent in the governing differential problem.

$$(2) \quad u_t + uu_x = 0, \quad u(x, 0) = u_0(x) = \begin{cases} 1 \text{ at } x \leq 0, \\ 0 \text{ at } x > 0. \end{cases}$$

Let us consider the Lax scheme:

$$u_j^{n+1} = \frac{1}{2}(u_{j+1}^n + u_{j-1}^n) - (\tau u_j^n/2h)(u_{j+1}^n - u_{j-1}^n)$$

the first differential approximation of which is $u_t + uu_x + \dfrac{h^2}{6}uu_{xxx} =$

$$= \frac{1}{2}\left(\frac{1}{k} - u^2 k\right)hu_{xx} - khuu_x^2.$$ Here the term $(h^2/6)\,uu_{xxx}$ is the local numerical dispersion, $(h/2k)(1 - u^2k^2)$ is the alternating coefficient of numerical viscosity (it is positive, since for stability the calculations should be carried out at $u\tau/h < 1$, i.e. $uk < 1$). The Lax scheme shows smoothing of the solutions.

For the Lax–Wendroff scheme, where

$$u_{j+\frac{1}{2}}^{n+\frac{1}{2}} = \frac{1}{2}(u_{j+1}^n + u_j^n) - \frac{\tau}{4h}[(u_{j+1}^n)^2 - (u_j^n)^2] \quad \text{is a predictor,}$$

$$u_j^{n+1} = u_j^n - \frac{\tau}{2h}[(u_{j+\frac{1}{2}}^{n+1/2})^2 - (u_{j-\frac{1}{2}}^{n+\frac{1}{2}})^2] \quad \text{is a corrector,}$$

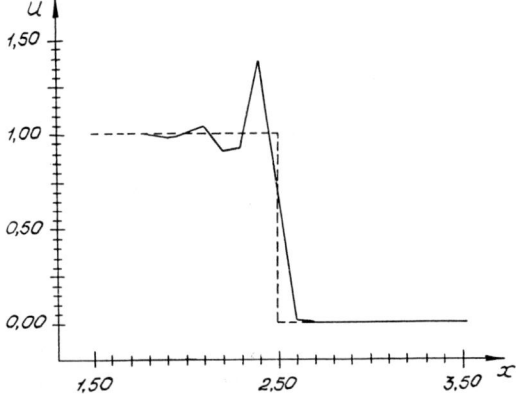

Fig. 1. Solution of the quasilinear transport equation. Dashed line shows the exact solution, solid line is the numerical solution by the Lax–Wendroff scheme ($\tau = 0.05$, $h = 0.1$).

the first differential approximation is cumbersome. The numerical results of calculations by this scheme are given in Fig. 1. The numerical solution obtained by the McCormack scheme

$$\bar{u}_j = u_j^n - \frac{\tau}{2h} [(u_{j+1}^n)^2 - (u_j^n)^2],$$

$$u_j^{n+1} = \frac{1}{2} (\bar{u}_j + u_j^n) - \frac{\tau}{4h} (\bar{u}_j^2 - \bar{u}_{j-1}^2)$$

is shown in Fig. 2.

These difference schemes distort the monotonic solutions, since they have numerical dispersion. Despite this, they can be used (having, naturally, been modified) when solving the KdV and BKV equations. In this case attention should be paid to the fact that the numerical dispersion be substantially less than the physical one, which can be regulated by the step h along the spatial coordinate, since the numerical dispersion is proportional to h.

In spite of the created analytical theory [51], allowing one to find, in principle, unsteady solutions of the KdV equation through the use

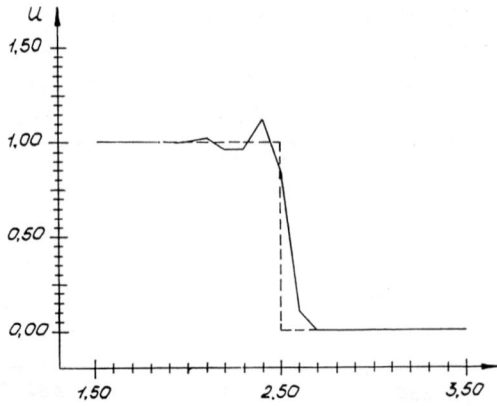

Fig. 2. Solution of the quasi-linear transport equation. Dashed line shows the exact solution, solid line is the numerical solution by the McCormack scheme ($\tau = 0.05$, $h = 0.1$).

of the inverse problem of the theory of scattering, the problem of numerical algorithms for the KdV and similar equations seems to be important, as the solution in an explicit analytical form can hardly be obtained at arbitrary initial data.

Let us consider some numerical algorithms for the KdV equation written as (2.1) [10].

1. The three-level explicit scheme with the order of approximation $0(\tau^2, h^2)$:

$$u_j^{n+1} = u_j^{n-1} - (u_j^n \tau/h)(u_{j+1}^n - u_{j-1}^n)$$
$$- (\beta\tau/h^3)(u_{j+2}^n - 2u_{j+1}^n + 2u_{j-1}^n - u_{j-2}^n). \tag{2.13}$$

The above scheme was used to get the first numerical KdV solutions [18, 111]. According to [10], scheme (2.13) is stable when the condition

$$(\tau/h)(|u| + 3\sqrt{3}\,\beta/2h^2) \leq 1$$

is met, or at sufficiently small steps h

$$\tau \leq 2h^3/3\sqrt{3}\,\beta \approx 0.384h^3/\beta. \tag{2.14}$$

As has been demonstrated by calculations, this condition is fairly exact. For instance, at $\beta = 2 \times 10^{-4}$, $h = 10^{-2}$ the calculation by scheme (2.13) was stable at $\tau = 1.91 \times 10^{-3}$ and unstable at $\tau = 1.93 \times 10^{-3}$ (formula (2.14) gives $\tau_{max} = 1.92 \times 10^{-3}$. Note, that the stability criterion of the scheme considered $(\tau/h)(|u| + 4\beta/h^2) \le 1$, given in [106], is wrong.

2. The two-step explicit Lax–Wendroff scheme:

$$u_{j+1/2}^{n+1/2} = \frac{1}{2}(u_{j+1}^n + u_j^n) - \frac{\tau}{4h}\left[(u_{j+1}^n)^2 - (u_j^n)^2\right.$$

$$\left. + \frac{2\beta}{h^2}(u_{j+2}^n - 3u_{j+1}^n + 3u_j^n - u_{j-1}^n)\right],$$

$$u_j^{n+1} = u_j^n - \frac{\tau}{2h}\left[(u_{j+1/2}^{n+1/2})^2 - (u_{j-1/2}^{n+1/2})^2\right.$$

$$\left. + \frac{2\beta}{h^2}(u_{j+3/2}^{n+1/2} - 3u_{j+1/2}^{n+1/2} + 3u_{j-1/2}^{n+1/2} - u_{j-3/2}^{n+1/2})\right]. \tag{2.15}$$

The condition of stability $\tau \le 0.25 h^3/\beta$ is more restricted than that for scheme (2.13); the approximation is $0(h^2)$.

3. The three-level explicit scheme with the order of approximation $0(\tau^2, h^4)$:

$$u_j^{n+1} = u_j^{n-1} + (u_j\tau/6h)(u_{j+2} - 8u_{j+1}$$

$$+ 8u_{j-1} - u_{j-2}) + (\beta\tau/4h^3)(u_{j+3} \tag{2.16}$$

$$- 8u_{j+2} + 13u_{j+1} - 13u_{j-1} + 8u_{j-2} - u_{j-3}), \quad u \equiv u^n.$$

In line with [10], the above scheme is stable when at sufficiently small steps h the condition

$$\tau \le \frac{108h^3}{(43 + 7\sqrt{73})\sqrt{10\sqrt{73} - 62}\,\beta} \approx 0.216 h^3/\beta \tag{2.17}$$

is met. Comparing formulae (2.14), (2.17) demonstrates that for a scheme with a higher order of approximation the condition of stability is more restrictive.

4. The two-level implicit scheme:

$$u_j^{n+1} = u_j - (u_j\tau/24h)(u_{j-2}^{n+1} - 8u_{j-1}^{n+1} + 8u_{j+1}^{n+1}$$

$$- u_{j+2}^{n+1}) - (u_{j+1}^{n+1}\tau/24h)(u_{j-2} - 8u_{j-1}\lambda + 8u_{j+1}$$

$$- u_{j+2}) + (\beta\tau/16h^3)(u_{j+3}^{n+1} - 8u_{j+2}^{n+1} + 13u_{j+1}^{n+1} \qquad (2.18)$$

$$- 13u_{j-1}^{n+1} + 8u_{j-2}^{n+1} - u_{j-3}^{n+1} + u_{j+3} - 8u_{j+2} + 13u_{j+1}$$

$$- 13u_{j-1} + 8u_{j-2} - u_{j-3}), \qquad u \equiv u^n.$$

The order of approximation is $0(\tau^2, h^4)$, the realization is a seven-point sweeping, the stability is absolute.

In calculations it is important to have estimates of the phase shift, which can be obtained through difference schemes. Let us consider scheme (2.16). If we fix the coefficient $u_j^n \equiv u$, then, making use of some initial data, we can get the solution

$$u_j^n = \sum_k a_k \frac{1 + \cos\theta}{2\cos\theta} e^{-i(\theta n + kjh)}$$

$$+ \sum_k (-1)^n a_k \frac{1 - \cos\theta}{2\cos\theta} e^{i(\theta n - kjh)}, \qquad (2.19)$$

where the θ value is determined by the expression

$$\theta = \arcsin\left\{\frac{\tau}{h}\left[\left(1 + \frac{2}{3}\sin^2\frac{kh}{2}\right)u\right.\right.$$

$$\left.\left. - \frac{4\beta}{h^2}\sin^2\frac{kh}{2}\cdot\left(1 + \sin^2\frac{kh}{2}\right)\right]\right\}$$

and at $\tau, h \to 0$ tends to $\omega_k\tau = (u - \beta k^2)k\tau$ (after cancelling to τ the latter is the law of dispersion of the KdV equation).

The second sum in expression (2.19) for u_j^n is a 'numerical' wave with its amplitude proportional to the step τ, while the first sum at $\tau, h \to 0$ gives a true solution. At the finite values of τ, h the phase velocity of this wave is determined by the formula

$$v_{ph} = (k\tau)^{-1} \arcsin \left\{ \frac{\tau}{h} \left[\left(1 + \frac{2}{3} \sin^2 \frac{kh}{2} \right) u \right. \right.$$
$$\left. \left. - (4\beta/h^2) \left(1 + \sin^2 \frac{kh}{2} \right) \sin^2 \frac{kh}{2} \right] \sin kh \right\}.$$

As shown in [106], the phase velocity for scheme (2.13) with the order of approximation $0(\tau^2, h^2)$ is

$$v_{ph} = (k\tau)^{-1} \arcsin \left[\frac{\tau}{h} \left(u - \frac{4\beta}{h^2} \sin^2 \frac{kh}{2} \right) \sin kh \right].$$

A phase shift between the approximate (at finite τ, h) and exact (at τ, $h \to 0$) solutions is defined by the formula

$$\Delta = 2\pi \left(1 - \frac{v_{ph}}{\omega_k/k} \right).$$

At sufficiently small steps τ, h, the phase shift for scheme (2.16) is

$$\Delta = 2\pi \left\{ 1 + (\omega_k\tau)^{-1} \arcsin \left[\frac{4\beta\tau}{h^2} \left(1 + \sin^2 \frac{kh}{2} \right) \sin^2 \frac{kh}{2} \sin kh \right] \right\},$$

and for scheme (2.13) is

$$\Delta = 2\pi \left\{ 1 + (\omega_k\tau)^{-1} \arcsin \left(\frac{4\beta\tau}{h^2} \sin^2 \frac{kh}{2} \sin kh \right) \right\}.$$

Comparing these formulae reveals that for the scheme with the order of approximation $0(\tau^2, h^4)$ the phase shift is greater than that for the scheme with the order of approximation $0(\tau^2, h^2)$.

The three-level schemes (2.13), (2.16) are rather simple, fairly accurate and convenient for realization. However, one has to remember more quantities at every time level than in two-level schemes, as well as to calculate the sought function at the moment $t = \tau$ by another algorithm; these drawbacks, though, being of minor importance due to the one-dimensionality of the problem. The calculations, presented in [10], have demonstrated a good quality of three-level schemes for obtaining the KdV equation solution.

The solution of the BKV equation (2.11) was carried out by the author of [10] by the following absolutely stable implicit scheme with the order of approximation $O(\tau^2, h^4)$:

$$u_j^{n+1} = u_j - \frac{\tau}{4h} [u_j(u_{j-2}^{n+1} - 8u_{j-1}^{n+1} + 8u_{j+1}^{n+1} - u_{j+2}^{n+1})$$

$$+ u_j^{n+1} (u_{j-2} - 8u_{j-1} + 8u_{j+1} - u_{j+2})]$$

$$+ \frac{\beta\tau}{16h^3} (u_{j+3}^{n+1} - 8u_{j+2}^{n+1} + 13u_{j+1}^{n+1} - 13u_{j-1}^{n+1} + 8u_{j-2}^{n+1} - u_{j-3}^{n+1}$$

$$+ u_{j+3} - 8u_{j+2} + 13u_{j+1} - 13u_{j-1} + 8u_{j-2} - u_{j-3})$$

$$- \frac{\alpha\tau}{24h^2} (u_{j-2}^{n+1} - 16u_{j-1}^{n+1} + 30u_j^{n+1} - 16u_{j+1}^{n+1} + u_{j+2}^{n+1}$$

$$+ u_{j-2} - 16u_{j-1} + 30u_j - 16u_{j+1} + u_{j+2}),$$

$$u \equiv u_j^n,$$

which ensures an economic way of obtaining calculation results of high accuracy.

To solve the KdV and BKV equations, the semi-spectral [38], as well as the spectral [1] methods have been developed. The first group of methods includes the so-called ASD-method (accurate space derivative method) that uses the expansion

$$u(t + \tau, x) = u(t, x) + \tau \frac{\partial u(t, x)}{\partial t} + \frac{\tau^2}{2} \frac{\partial^2 u(t, x)}{\partial t^2} + \frac{\tau^3}{6} \frac{\partial^3 u(t, x)}{\partial t^3}$$

for obtaining the sought function at the subsequent moment of time, and employs equation (2.11) through the formulae

$$\frac{\partial u}{\partial t} = - u \frac{\partial u}{\partial x} + \alpha \frac{\partial^2 u}{\partial x^2} - \beta \frac{\partial^3 u}{\partial x^3},$$

$$\frac{\partial^2 u}{\partial t^2} = - \frac{\partial u}{\partial t} \frac{\partial u}{\partial x} - \left(u \frac{\partial}{\partial x} - \alpha \frac{\partial^2}{\partial x^2} + \beta \frac{\partial^3}{\partial x^3} \right) \frac{\partial u}{\partial t},$$

$$\frac{\partial^3 u}{\partial t^3} = - \frac{\partial^2 u}{\partial t^2} \frac{\partial u}{\partial x} - 2 \frac{\partial u}{\partial t} \frac{\partial^2 u}{\partial t \partial x} - \left(u \frac{\partial}{\partial x} - \alpha \frac{\partial^2}{\partial x^2} + \beta \frac{\partial^3}{\partial x^3} \right) \frac{\partial^2 u}{\partial t^2}.$$

for calculating the time derivatives at the moment of time t. The derivatives with respect to the coordinate x are calculated by the Fourier method. If $U(k, t)$ is a Fourier-transform of the function $u(x, t)$ on the introduced grid, then the derivative of, for instance, the m th order is determined by the formula $\partial^m u / \partial x^m = \sum_k (ik)^m U(k, t) e^{ikx}$, where the summation is carried out over all the wave numbers k that can be found on the grid covering the domain of the numerical solution.

Within the spectral method one carries out the expansion of the sought solution of equation (2.11) in a series, setting the periodic boundary conditions, $\sum_{k=-\infty}^{\infty} a_k(t) \exp (ikx) = u(x, t)$. Substituting this expansion into the governing KdV equation, we get the following infinite system of equations for the amplitudes of the harmonics:

$$\frac{da_k}{dt} = -\frac{1}{2} ik \sum_{m=-\infty}^{\infty} a_{k-m} a_m + i\beta k^3 a_k.$$

In order to make this system finite, we introduce the maximum wave number $k_{max}(a_k(t) = 0$ for all $|k| > k_{max}$), the value of which is chosen with the help of trial calculations with growing k_{max}. These methods proved to be fairly exact [1, 38], but in a technical sense they are much more complex than the finite-difference methods, and are inefficient as they consume a lot of computing time.

2.2. Auto-Oscillations in an Anisotropic Plasma

Let us consider the Alfven waves of a small but finite amplitude, propagating along the strong magnetic field $B_0 = \{0, 0, B_0\}$ in a rarefied plasma [6, 30]. For this purpose let us assume that in the system of equation (1.36) $B_\perp^2 \ll 1$, $B_\perp^2 = (B_x^2 + B_y^2)/B_0^2$. In this case the last two equations of this system give

$$p_\| \approx p_\|^0 (1 - B_\perp^2 + B_\perp^4), \quad p_\perp \approx p_\perp^0 \left(1 + \frac{1}{2} B_\perp^2 - \frac{1}{8} B_\perp^4\right) \quad (2.20)$$

where $p_\parallel^0$ is the initial longitudinal pressure of the plasma, $p_\perp^0$ is the initial transversal pressure of the plasma. Let us substitute expressions (2.20) into system (1.36) and neglect the terms $B_\perp^4$; let us also assume that the value of the Larmor frequency Ω_B is sufficiently great and in the brackets after the coefficient Ω_B^{-1} neglect the terms that are proportional to $B_\perp^2$, $B_\perp^4$. As a result, we get the following system of equations:

$$\varrho\frac{\partial u}{\partial t} + \frac{\partial}{\partial z}\left\{\left[\Delta p - \left(2p_\parallel^0 - \frac{1}{2}p_\perp^0\right)B_\perp^2\right]H - \Omega_B^{-1}p_\parallel^0\frac{\partial v}{\partial z}\right\} = 0,$$

$$\varrho\frac{\partial v}{\partial t} + \frac{\partial}{\partial z}\left\{\left[\Delta p - \left(2p_\parallel^0 - \frac{1}{2}p_\perp^0\right)B_\perp^2\right]B + \Omega_B^{-1}p_\parallel^0\frac{\partial u}{\partial z}\right\} = 0, \quad (2.21)$$

$$\frac{\partial H}{\partial t} = \frac{\partial u}{\partial z}, \qquad \frac{\partial B}{\partial t} = \frac{\partial v}{\partial z}.$$

Here $\Delta p = p_\parallel^0 - p_\perp^0 - B_0^2/\mu_0$ is the degree of the plasma anisotropy, $H = B_x/B_0$, $B = B_y/B_0$. It can be easily proved that by way of substituting the variables in system (2.21) we can get rid of the coefficients $p_\parallel^0$, $p_\perp^0$, Ω_B related to a concrete condition of the plasma. Indeed, let

$$t' = (\Omega_B\Delta p/p_\parallel^0)\,t, \qquad z' = (\Omega_B/p_\parallel^0)\,(\varrho\,\Delta p)^{1/2}\,z,$$

$$H', B' = \left[\left(2p_\parallel^0 - \frac{1}{2}p_\perp^0\right)\Big/\Delta p\right]^{1/2}H, B, \quad (2.22)$$

$$u', v' = \left[\varrho\left(2p_\parallel^0 - \frac{1}{2}p_\perp^0\right)\right]^{1/2}(\Delta p)^{-1}\,u, v$$

then in these variables the system of equations (2.21) assumes the following form (primes omitted):

$$\frac{\partial u}{\partial t} + \frac{\partial}{\partial z}\left[(1 - B_\perp^2)\,H - \frac{\partial v}{\partial z}\right] = 0,$$

$$\frac{\partial v}{\partial t} + \frac{\partial}{\partial z}\left[(1 - B_\perp^2)\,B + \frac{\partial u}{\partial z}\right] = 0, \quad (2.23)$$

$$\frac{\partial H}{\partial t} = \frac{\partial u}{\partial z}, \qquad \frac{\partial B}{\partial t} = \frac{\partial v}{\partial z}.$$

System (2.23) has, thus, no coefficients and, hence, the solutions of the problem of the development of the firehose instability of the Alfven waves in a rarefied anisotropic plasma for various parameters characterizing the initial state of the plasma, will be similar in the approximation of small but finite magnetic fields $B_\perp$. The system of equations (2.23), being a convenient model for studying the instability in question, is also of mathematical interest. Let us write the law of the energy conservation for system (2.23) setting, as in Chapter 1, that the problem is solvable on the interval $0 \div L$ with the periodical boundary conditions

$$\int_0^L Q_1(z, t)\, dz = \int_0^L Q_1(z, 0)\, dz,$$

where

$$Q_1(z, t) = \frac{1}{2}(u^2 + v^2) + \frac{1}{4}(B_\perp^2 - 1)^2.$$

Now let us investigate the model system (2.23). Let us first of all note that within this approximation the increment of the growth of small disturbances is $\gamma_k = k\left(1 - \frac{1}{4}k^2\right)^{1/2}$, the most rapidly increasing harmonic is that with the wave number $k = 2^{1/2}$ and for it $\gamma_k = 1$; the harmonics with $k < 2$ are unstable, while those with $k > 2$ are stable. System (2.23) has a partial solution that can be studied analytically, it is a monochromatic wave with a circular polarization

$$H(z, t) = A(t) \sin\left(kz + \varphi(t)\right), \qquad B(z, t) = A(t) \cos\left(kz + \varphi(t)\right).$$

$$(2.24)$$

In such a wave, as can be easily seen, the square of the perpendicular magnetic field is only the function of time: $B_\perp^2 = H^2 + B^2 = A^2(t)$. Therefore, substituting the required form of the solution (2.24) into system (2.23) we come to the conclusion that the velocity of the phase variation (frequency) of the wave is $\dot{\varphi}(t) = -\dfrac{1}{2}k^2 + CA^{-2}(t)$,

$C = \left(\dot{\varphi}(0) + \dfrac{1}{2}k^2\right)A^2(0)$ is the constant of the integration, and the wave amplitude $A(t)$ is determined by the ordinary differential equation of the non-linear oscillator

$$\ddot{A} - \gamma_k^2 A + k^2 A^3 - C^2 A^{-3} = 0, \qquad (2.25)$$

where $\dot{A}^2 + U(A) = E = \text{const.}$, the 'potential' energy is $U(A) = \dfrac{1}{2}k^2 A^2 \left(A^2 + \dfrac{1}{2}k^2 - 2\right) + C^2 A^{-2}$, while the 'total' energy is $E = \dot{A}^2(0) + \dfrac{1}{2}k^2\left(A^2(0) + \dfrac{1}{2}k^2 - 2\right)A^2(0) + C^2 A^{-2}(0)$, $\dot{A}(0) = \left.\dfrac{\mathrm{d}A}{\mathrm{d}t}\right|_{t=0}$. Analysis of the function U makes it possible to present the qualitative changes in the wave amplitude in the following way. Let at the initial moment $t = 0$ there be an Alfven wave of a small amplitude $A(0) \ll 1$ with the wave number k, lying in the domain of instability $k < 2$. At comparatively small times the amplitude of this wave will grow exponentially due to instability until it reaches the value when the non-linear terms gain in importance. Growth of the amplitude slows down, it reaches its maximum, then there occurs a decrease up to the minimal value, then an increase again and so on — there arise periodical oscillations of the amplitude with respect to its value with growing time. If in (2.25) we set the integration constant C equal to zero and that $\dot{A}(0) = \gamma_k A(0)$, then

$$A_{\min} = 0, \quad A_{\max} = B_{\perp \max} = \left[2\left(1 - \dfrac{1}{4}k^2\right)\right]^{1/2}. \qquad (2.26)$$

It results from formula (2.26) that the maximum square of a perpendicular magnetic field in a non-linear wave reduces with growing wavelength and turns, naturally, to zero on the boundary of instability $k = 2$. At sufficiently great values of the integration constant

C we can find the asymptotic expression $A_{max} \sim C^{1/2}$. If we go over to dimensional variables through formulae (2.22), then, in the case when $C = 0$, $\dot{\varphi}(0) = -\dfrac{1}{2} \Omega_B kR$, we get

$$B_\perp^2(t) = \frac{B_0^2 \, \Delta p}{2p_\parallel^0 - \dfrac{1}{2} p_\perp^0} A^2(t),$$

$$A_{max}^2 = 2\left(1 - \frac{p_\parallel^0}{4\Delta p} k^2 R^2\right),$$

$$B_{\perp max}^2 / B_0^2 = \frac{\gamma_k^2}{(\Omega_B kR)^2 \, (1 - p_\perp^0/4p_\parallel^0)}, \qquad (2.27)$$

$$\gamma_k = \Omega_B kR \left(\Delta p/p_\parallel^0 - \frac{1}{4} k^2 R^2\right)^{1/2}.$$

If we assume $p_\parallel^0 \sim p_\perp^0 \gg B_0^2/\mu_0$, then in the long-wave limit $(kR \to 0)$ we have

$$B_{\perp max}^2 / B_0^2 \to \frac{\Delta p}{p_\parallel^0 - \dfrac{1}{4} p_\perp^0} \approx 4\Delta p/3p_\parallel^0.$$

In the conditions of changing the amplitude of the Alfven wave passing through the plasma, the gas-kinetic pressures do not remain constant with growing time; neglecting the quantities $B_\perp^4$ we get from formulae (2.20)

$$p_\parallel(t) = p_\parallel^0 \left(1 - \frac{\Delta p}{2p_\parallel^0 - \dfrac{1}{2} p_\perp^0} \cdot A^2(t),\right)$$

$$p_\perp(t) = p_\perp^0 \left(1 + \frac{\Delta p}{4p_\parallel^0 - p_\perp^0} \cdot A^2(t)\right). \qquad (2.28)$$

If the wave amplitude grows, the longitudinal pressure of the plasma decreases, while the transversal pressure increases, which results in the reduction of the initial degree of the plasma anisotropy:

$$\Delta p(t) = p_{\parallel}(t) - p_{\perp}(t) - \left(B_0^2 + B_{\perp}^2(t)\right)/\mu_0$$

$$= \left(1 - \frac{p_{\parallel}^0 + \dfrac{1}{2}\, p_{\perp}^0 + B_0^2/\mu_0}{2p_{\parallel}^0 - \dfrac{1}{2}\, p_{\perp}^0} \cdot A^2(t)\right) \Delta p.$$

When the magnetic field in a wave reaches its maximum, the anisotropy degree becomes equal to

$$\Delta p(B_{\perp \max}) \equiv \Delta p_1 =$$

$$= \left[1 - \frac{p_{\parallel}^0 + \dfrac{1}{2}\, p_{\perp}^0 + B_0^2/\mu_0}{p_{\parallel}^0 - \dfrac{1}{4}\, p_{\perp}^0} \cdot \left(1 - (p_{\parallel}^0/4\Delta p)\, k^2 R^2\right)\right] \Delta p. \qquad (2.29)$$

In the long-wave approximation $kR \to 0$ and for the plasma with great values of gas-kinetic pressure $p_{\parallel}^0 \sim p_{\perp}^0 \gg B_0^2/\mu_0$, it results from formula (2.29) that $\Delta p_1 = -\Delta p$, therefore, the anisotropy $\Delta p(t)$ changes in the wave periodically from Δp to $-\Delta p$, passing through zero. On the boundary of instability, when $(kR)^2 = 4\Delta p/p_{\parallel}^0$, the $\Delta p(t)$ value does not change (since instability is not excited). Finally, if in the plasma a wave with a maximum increment is propagating (it should be recalled that $\gamma_k = \gamma_{\max}$ for the wave with $(kR)^2 = 2\Delta p/p_{\parallel}^0$), then $\Delta p_1 = \dfrac{1}{2}\,(\Delta p)^2 \Big/ \left(p_{\parallel}^0 - \dfrac{1}{4}\, p_{\perp}^0\right) \approx 2(\Delta p)^2/3p_{\parallel}^0 > 0$.

Writing the solution of equation (2.25) through elliptical functions, one can find an approximate formula for the period of change in the squared amplitude of a non-linear Alfven wave for the case when

$$C = 0, \quad \dot{\varphi}(0) = -\frac{1}{2}\,\Omega_B k R, \quad \dot{A}(0) = \gamma_k A(0) \; [7]:$$

$$T_k \approx \frac{2}{\gamma_k} \ln\left[8A^{-2}(0) \cdot \left(1 - \frac{p_{\parallel}^0}{4\Delta p}\, k^2 R^2\right)\right] \qquad (2.30)$$

In the long-wave approximation $kR \to \infty$ we have

$$\gamma_k \approx \frac{2}{k}(\varrho/\Delta p)^{1/2}, \quad T_k \approx (\Omega_B kR)^{-1}(p_\parallel^0/\Delta p)^{1/2}\ln(8A^{-2}(0)) \sim (kR)^{-1}.$$

$$(2.31)$$

The period of oscillations of the velocities and magnetic fields in a wave is determined by the conventional formula $T^{(k)} = 2\pi/\omega^{(k)}$, or, using (1.37), we get

$$T^{(k)} = 4\pi\Omega_B^{-1}(kR)^{-2} \sim (kR)^{-2}. \qquad (2.32)$$

Comparing formulae (2.31) and (2.32) shows that in the case of long waves $T^{(k)} \gg T_k$. For the wave with the wave number $k = (2\Delta p/p_\parallel^0)^{1/2} R^{-1}$ and with the maximum increment $\gamma_{\max} = \Omega_B \Delta p/p_\parallel^0$, the period ratio is $T_k/T^{(k)} = \frac{2}{\pi}\ln(2A_0^{-1})$, the period T_k can be greater than $T^{(k)}$.

The quasi-linear theory (QLT) of firehose instability has been considered in [63, 98]. Comparison of the macroscopic plasma characteristics, carried out on the basis of the QLT and the Chew–Goldberger–Low (CGL) model at small anisotropy, has resulted in the following conclusions. In line with the QLT, the relation between changes in the longitudinal and transversal pressures is given by the formula

$$\frac{dp_\parallel}{dp_\perp} \approx -4\left(p_\parallel^0 - \frac{1}{2}p_\perp^0\right)\Big/ p_\parallel^0,$$

while according to the CGL model we have

$$\Delta p_\parallel \equiv p_\parallel(t) - p_\parallel^0 = -p_\parallel^0 B_\perp^2(t)/B_0^2,$$

$$\Delta p_\perp \equiv p_\perp(t) - p_\perp^0 = \frac{1}{2}p_\perp^0 B_\perp^2(t)/B_0^2,$$

$$\frac{\Delta p_\parallel}{\Delta p_\perp} \approx \frac{dp_\parallel}{dp_\perp} \approx -2p_\parallel^0/p_\perp^0.$$

Since for the QLT applicability, the conditions $p_\parallel^0$, $p_\perp^0 \gg B_0^2/\mu_0$, $(p_\parallel^0 - p_\perp^0)/p_\parallel^0 \ll 1$ must be met, both formulae can be reduced to one: $dp_\parallel/dp_\perp \approx -2$. The same is true for time changes of the plasma anisotropy $\Delta p(t) = p_\parallel(t) - p_\perp(t)$. Indeed, according to the quasi-linear theory

$$\Delta p(t) \approx \Delta p - (5p_\parallel^0 - 2p_\perp^0)\, h(t),$$

while by the CGL model we have

$$\Delta p(t) \approx \Delta p - (2p_\parallel^0 + p_\perp^0)\, h(t)$$

where $h(t) = \dfrac{1}{2} \sum_k B_{\perp k}^2/B_0^2$. Under the conditions of the QLT applicability both formulae afford

$$\Delta p(t) \approx \Delta p - 3p_\parallel^0 h(t).$$

Let us consider the effect exerted on the non-linear monochromatic wave by other waves, for which we set $k \ll k_0$ (k_0 is the wave number of the 'basic' wave). For these additional waves the increment γ_k will be much less than the increment $\gamma_{k_0} \equiv \gamma_0$, therefore, within the period of time of the order of the period T_0 of the basic wave, they can be considered in the linear approximation, i.e.

$$\frac{dh}{dt} = \sum_k 2\gamma_k h_k.$$

In this equation the increment γ_k will be a time function, since the pressures $p_\parallel$ and $p_\perp$, included in the expression for the increment, are related to the changes in the magnetic field in the basic wave.

If, for the sake of simplicity, we take an additional wave and the basic one with the wave number $k_0 = (2\Delta p/p_\parallel^0 R^2)^{1/2}$ we shall get the following estimate [6]: by the end of the period $T_{k_0} \equiv T_0$ of the basic wave, due to the growth of the additional wave amplitude, the square of the transversal magnetic field will become equal to $B_\perp^2(T_0) \approx (1 + \alpha k)\, B_{\perp k}^2(0)$, where α is a numerical coefficient. As a result, there arises irreversibility of changes in the wave amplitude and plasma characteristics, the degree of anisotropy decreases, the tendency of levelling the transversal and longitudinal pressures is observed, which in fact is in a qualitative agreement with the conclusion on the quasi-linear stabilization of firehose instability [63, 98].

The analytical solution (2.27), (2.28), describing a non-linear Alfven wave with a circular polarization, corresponds to the initial conditions of a quite definite type, i.e. (setting, for simplicity, the integration constant $C = 0$):

$$H(z, 0) = A(0) \sin kz, \quad B(z, 0) = A(0) \cos kz,$$

$$u(z, 0) = \frac{1}{k} [A(0)\, \dot{\varphi}(0) \sin kz - \dot{A}(0) \cos kz], \qquad (2.33)$$

$$v(z, 0) = \frac{1}{k} [A(0)\, \dot{\varphi}(0) \cos kz + \dot{A}(0) \sin kz].$$

At arbitrary initial conditions, a solution of the model system (2.23) can be found, in all probability, only with numerical methods. Out of a number of the algorithms that have been used [30] we are going to dwell upon two: (1) the algorithm based on spectral expansions, (2) the finite-difference iteration algorithm.

The spectral method
When considering the system of equations (2.23) within the domain $\{0 \le z \le L, 0 < t \le T\}$ with periodical boundary conditions and arbitrary, generally speaking, initial conditions, let us choose as a basis the complete system of the functions exp $(imkz)$, $k = 2\pi/L$, $m = 0, \pm1, \ldots$, obeying the periodical boundary conditions and orthogonal in the intercept $0 \le z \le L$. Use made of the Fourier transform results in an infinite system of ordinary differential equations for the harmonic amplitudes:

$$\frac{du_m}{dt} = -imk H_m + k^2 m^2 v_m + imk \sum_{n+n'+n''=m} H_n (H_{n'} H_{n''} + B_{n'} B_{n''}),$$

$$\frac{dv_m}{dt} = -imk B_m - k^2 m^2 u_m + imk \sum_{n+n'+n''=m} B_n (H_{n'} H_{n''} + B_{n'} B_{n''}),$$

$$\frac{dH_m}{dt} = imk u_m, \quad \frac{dB_m}{dt} = imk v_m \qquad (2.34)$$

with the boundary conditions $f(z, 0) = \sum\limits_{n=-\infty}^{\infty} f_m(0) \exp(imkz)$. To convert the infinite system (2.34) into a finite one, let us introduce the maximum wave number Nk and assume $f_m(t) = 0$ for all $|m| > N$. The thus obtained system of ordinary differential equations of the Nth order can be solved by any high-order method, for instance, by the Adams or the Runge-Kutta methods. In order to overcome technical difficulties, associated with the uneconomical methods of calculating the sums of type $R_m = \sum\limits_{|n|<N} \sum\limits_{|n'|<N} \sum\limits_{|n''|<N} A_n B_{n'} C_{n''}$, $n + n' + n'' = m$, let us make use of the method generalization [83] for the case of triple non-local sums. These are the calculating formulae:

(a) Fourier-transform of the sets H_m, B_m

$$\hat{h}(j) = \sum_{m=-N}^{N-1} H_m \exp(imx_j), \quad \hat{b}(j) = \sum_{m=-N}^{N-1} B_m \exp(imx_j),$$

$$\tilde{h}(j) = \sum_{m=-N}^{N-1} H_m \exp(imx_{j+1/3}), \quad \tilde{b}(j) = \sum_{m=-N}^{N-1} B_m \exp(imx_{j+1/3}),$$

$$\bar{h}(j) = \sum_{m=-N}^{N-1} H_m \exp(imx_{j+2/3}), \quad \bar{b}(j) = \sum_{m=-N}^{N-1} B_m \exp(imx_{j+2/3});$$

$$H_{-N} = B_{-N} = 0, \quad xj = \pi_j/N,$$

$$x_{j+1/3} = \pi(j + 1/3)/N, \quad x_{j+2/3} = \pi(j + 2/3)/N,$$

$$j = 0, 1, \ldots, (2N - 1);$$

(b) image multiplication

$$\hat{p}(j) = \hat{h}(j) \left(\hat{h}^2(j) + \hat{b}^2(j)\right), \quad \hat{q}(j) = \hat{b}(j) \left(\hat{h}^2(j) + \hat{b}^2(j)\right),$$

$$\tilde{p}(j) = \tilde{h}(j) \left(\tilde{h}^2(j) + \tilde{b}^2(j)\right), \quad \tilde{q}(j) = \tilde{b}(j) \left(\tilde{h}^2(j) + \tilde{b}^2(j)\right),$$

$$\bar{p}(j) = \bar{h}(j) \left(\bar{h}^2(j) + \bar{b}^2(j)\right), \quad \bar{q}(j) = \bar{b}(j) \left(\bar{h}^2(j) + \bar{b}^2(j)\right);$$

(c) inverse Fourier-transform

$$\hat{P}_m = (2N)^{-1} \sum_{j=0}^{2N-1} \hat{p}(j) \exp(-imx_j),$$

$$\hat{Q}_m = (2N)^{-1} \sum_{j=0}^{2N-1} \hat{q}(j) \exp(-imx_j);$$

the expressions $\tilde{P}_m$, $\tilde{Q}_m$, $\bar{P}_m$, $\bar{Q}_m$ are determined in an analogous way;

(d) triple sum calculations

$$P_m = \frac{1}{3} [\hat{P}_m + \tilde{P}_m \exp(-i\pi m/3N) + \bar{P}_m \exp(-2i\pi m/3N)],$$

$$Q_m = \frac{1}{3} [\hat{Q}_m + \tilde{Q}_m \exp(-i\pi m/3N) + \bar{Q}_m \exp(-2i\pi m/3N)],$$

where

$$P_m = \sum_{|n|<N} \sum_{|n'|<N} \sum_{|n''|<N} H_n(H_{n'}, H_{n''} + B_{n'}B_{n''}),$$

$$Q_m = \sum_{|n|<N} \sum_{|n'|<N} \sum_{|n''|<N} B_n(H_{n'}H_{n''} + B_{n'}B_{n''}),$$

$$n + n' + n'' = m.$$

In the algorithm in question we have to calculate the quantities of type $\hat{a}(j) = \sum_{m=-N}^{N-1} A_m w^{jm}$, where $w = \exp(i\pi/N)$, $j = 0, 1, \ldots, 2N-1$, in which case $4N^2$ of multiplications is required. The number of operations can be essentially reduced if we use the fast Fourier-transform [83]. The advantage being maximum when the number N is the degree of 2. In our case, using the fast Fourier-transform and choosing $N = 2^r$, we have only $16(3r + 5)N$ multiplications instead of $4(2N - 1)^2$ in calculating the quantities P_m, Q_m.

The finite-difference iteration algorithm

Let us consider the iteration scheme:

$$u_j^{n+\frac{1}{2},s+1} = u_j^n - \frac{\tau}{2}\left\{\Lambda_1\left(\frac{a^n + a^{n+1,s}}{2}H^{n+\frac{1}{2},s+1}\right)_j - \Lambda_{11}v_j^{n+\frac{1}{2},s+1}\right\},$$

$$v_j^{n+\frac{1}{2},s+1} = v_j^n - \frac{\tau}{2}\left\{\Lambda_1\left(\frac{a^n + a^{n+1\,s}}{2}B^{n+\frac{1}{2},s+1}\right)_j + \Lambda_{11}u_j^{n+\frac{1}{2},s+1}\right\},$$

$$H_{j+\frac{1}{2}}^{n+\frac{1}{2},s+1} = H_{j+\frac{1}{2}}^n + \frac{\tau}{2}\Lambda_1 u_{j+\frac{1}{2}}^{n+\frac{1}{2},s+1}, \qquad (2.35)$$

$$B_{j+\frac{1}{2}}^{n+\frac{1}{2},s+1} = B_{j+\frac{1}{2}}^n + \frac{\tau}{2}\Lambda_1 v_{j+\frac{1}{2}}^{n+\frac{1}{2},s+1},$$

$$u_j^{n+1,s+1} = 2u_j^{n+\frac{1}{2},s+1} - u_j^n, \qquad v_j^{n+1,s+1} = 2v_j^{n+\frac{1}{2},s+1} - v_j^n$$

and in an analogous way for the functions H^{n+1}, B^{n+1}. Here $a = 1 - H^2 - B^2$, $(\Lambda_1 f)_j = (f_{j+\frac{1}{2}} - f_{j-\frac{1}{2}})h^{-1}$, $(\Lambda_{11}f)_j = (f_{j+1} - 2f_j + f_{j-1})h^{-2}$, $f_{j+\frac{1}{2}} = \frac{1}{2}(f_j + f_{j+1})$, $j = 0, 1, \ldots, N$; s is the number of the iteration; as an initial approximation the function value at the preceding moment of time is chosen, i.e. $f^{n+1,0} = f^n$.

For scheme (2.35) the difference increment of instability is as follows:

$$\gamma_k(\tau, h) = \mathscr{H}\left(1 - \frac{1}{4}\mathscr{H}^2\right)^{1/2}\left[1 - \frac{1}{12}\mathscr{H}^2(\mathscr{H}^2 - 1)\tau^2\right]$$

where $\mathscr{H} = 2h^{-1}\sin\left(\dfrac{kh}{2}\right)$. Therefore, the domain of instability of the difference problem coincides with that of the differential problem to the accuracy of the terms of the order h^2, while the difference increment coincides with the differential one to the accuracy of $0(\tau^2, h^2)$. To estimate the qualitative properties of the scheme, the finite-difference solution has been compared with the exact solution (in the form of the circularly polarized Alfven wave, which is the solution of equation (2.25)). Under the exact solution we understand the solution of equation (2.25) for the wave amplitude $A(t)$, obtained by the Runge-Kutta method with the order of approximation $0(\tau^4)$. By scheme (2.35) we get the numerical solution for scheme (2.23), corresponding to the initial conditions (2.33), and compare the mean with respect to the coordinate z square of the magnetic field

$$\langle B_\perp^2(t^n) \rangle = N^{-1} \sum_{j=1}^{N} [(H_j^n)^2 + (B_j^n)^2]$$

with the value of $A^2(t^n) = H^2(t^n) + B^2(t^n)$. The difference analogue of the total energy is as follows

$$\mathscr{W}^n = h \sum_{j=1}^{N} \left\{ \frac{1}{2} [(u_j^n)^2 + (v_j^n)^2] + \frac{1}{4} [1 - (H_{j+\frac{1}{2}}^n)^2 - (B_{j+\frac{1}{2}}^n)^2] \right\}.$$

After the algebraic transformations with account taken of the period-ical boundary conditions, for the iteration scheme (2.35) under the condition of the iteration convergence we have $\mathscr{W}^{n+1} = \mathscr{W}^n$. There-fore the scheme being verified exhibits good qualities and can be used in calculations [30].

In order to compare the spectral and difference methods, calcula-tions have been carried out using the same parameters. Optimization of the spectral method (the use of the fast Fourier-transform) allows one to reduce the number of operations even in the case of a limited number of harmonics; for instance, if the number of harmonics is 2^5, the gain in time is 1.5. The iteration scheme, however, proves to be faster at the same degree of accuracy.

With the view of elucidating the evolution of single harmonics, as well as of sets of various harmonics with small initial amplitudes, the author has carried out numerical solutions of system (2.23) by the above iteration scheme. Let at the moment $t = 0$ the amplitudes of all the harmonics with the numbers $m \neq m_1$ equal zero, and the func-tions $(u, v, H, B)_{m_1}$ be arbitrary. In this case the energy of the har-monic with the number m_1 is transferred to the harmonics with the numbers $m = \pm(2l + 1) m_1$, $l = 0, 1, \ldots$ If conditions (2.33) are chosen as initial, the energy remains only in the harmonic with the number m_1. If at the initial moment of time the amplitudes of the harmonics with the numbers $m_1, m_2, \ldots, m_l$ are other than zero, the energy of these harmonics is given off to all the harmonics with the numbers nm_0, where m_0 is the greatest common divisor of the numbers $m_1, m_2, \ldots, m_l$; $n = \pm 1, \pm 2, \ldots$ With growing time this energy is redistributed among the same harmonics.

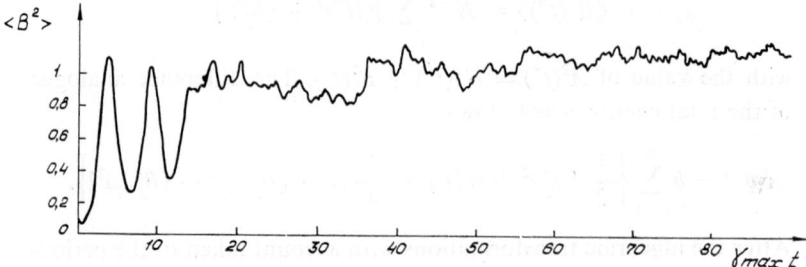

Fig. 3. Evolution of the mean square of the transverse magnetic field.

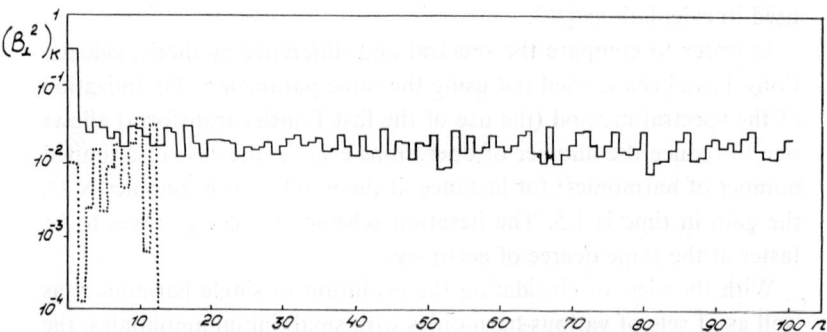

Fig. 4. Spectral density of the magnetic field energy (dashed line for $t = 0$, solid line for the moment $t = 100\,\gamma_{\max}^{-1}$).

If at the moment $t = 0$ we set a random distribution of the functions u, v, H, B with small amplitudes, then at first each harmonic, corresponding to the wave number k from the domain of instability, increases proportionally to exp $(\gamma_k t)$. When the harmonic amplitudes become sufficiently great, a non-linear interaction begins among them and, as a result, the quasi-stationary regime is established with growing time. The mean square of the transversal magnetic field in the wave

$\langle B_\perp^2 \rangle$ is affected by random oscillations at a certain level that is independent from the distribution of the given initial energy over the harmonics. As is seen from Fig. 3, which presents the time evolution of $\langle B_\perp^2(t) \rangle$, this average level is reached during the times $\approx 20\gamma_{max}^{-1}$. Figure 4 presents the density distribution of the magnetic field energy over the harmonics (dashed line — at the moment $t = 0$, solid line — at $t = 100_{max}^{-1}\gamma$). According to calculations, within a certain period of time the maximum energy will be concentrated in the first harmonic, having the greatest wavelength, while each of the other harmonics will have approximately 10% of this maximum value.

The whole system of equations (1.36), as has been proved, also has a partial analytical solution for arbitrary amplitudes in the form of a monochromatic wave with the circular polarization [9, 27] $B_x = B_0 A(t) \sin \left(kz + \varphi(t)\right)$, $B_y = B_0 A(t) \cos \left(kz + \varphi(t)\right)$, where the phase $\varphi(t)$ and the amplitude $A(t)$ obey the equations:

$$\dot\varphi = -\omega^{(k)} A^{-2} \left[1 + \frac{4}{3}\delta + \frac{2}{3}\delta(A^2 - 2) D^{1/2} - D^{-1} \right],$$

$$\dot{A}^2 + U(A) = E = \text{const.},$$

$$U(A) = (p_\parallel^0/\varrho) k^2 (D^{-1} + 2\delta D^{1/2} + B_0^2 A^2/4\pi p_\parallel^0)$$

$$+ (\omega^{(k)})^2 [(\delta/3)^2 (32 A^{-2} - 12 A^2 + 4 A^4 - 32 A^{-2} D^{1/2}$$

$$+ 16 D^{1/2}) + (4\delta/3) (D^{1/2} - 2 A^2 D^{-1} + 3 D^{-1/2})$$

$$- (A^2 + D) D^{-1} - A^4 D^{-2}],$$

$$\delta \equiv p_\perp^0/p_\parallel^0, \quad D \equiv 1 + A^2.$$

The total energy E is governed by the amplitude of the initial disturbance and is $E = \gamma_k^2 \dot{A}^2(0) + U(A_0)$. As was the case for the Alfven waves of a finite but small amplitude, described by the system of equations (2.23), the amplitude of the non-linear wave in question also periodically changes in time, the character of these oscillations being determined by the type of the function $U(A)$ and the value of the initial disturbance. As results from calculations, the maximum wave amplitude A_{max} increases with growing initial anisotropy.

By the iteration difference scheme, which is a generalization of scheme (2.35) for system (2.23), a major series of calculations has been carried out for the processes that are described by the complete system of equations (1.36). The initial conditions were chosen as follows: a monochromatic circularly polarized wave of a small amplitude (a solution of the linearized analogue of system (1.36)) and a random set of other waves with their amplitudes considerably less than the basic one.

The calculations carried out at various parameters of the initial state of the plasma gave, in a qualitative respect, the similar results [9, 27]: at comparatively small times ($t \lesssim 20 \div 30\gamma_0^{-1}$, where γ_0 is an increment of the basic excited harmonic) there take place non-linear regular oscillations in the plasma and field characteristics, at which the mean with respect to the calculation interval square of the transversal magnetic field first increases up to the values, showing good coincidence with the analytical solution, then $\langle B_\perp^2 \rangle$ 'goes back' nearly to zero, and then the process repeats itself. With growing time, due to interactions of the basic wave with the random given waves and the 'noises' of the difference scheme, regular oscillations of the magnetic field disappear and the $\langle B_\perp^2 \rangle$ value reaches the quasi-stationary level [9, 27], in which case the spatial distribution of the magnetic field and other characteristics of the system get stochastic. Therefore, subsequent to regular (laminar) oscillations the system gets under the turbulent conditions due to the effect of random disturbances. The value of the mean square of the magnetic field under these turbulent conditions increases with growing degree of the plasma anisotropy.

The calculation results indicate that at comparatively small times after the development of instability, all the magnetic energy is concentrated in the basic harmonic that was set as an initial condition. In the course of the process of getting into the turbulent state, when the oscillations become stochastic, the spectrum begins changing, since alongside with the basic harmonic there appear other harmonics. A greater portion of the magnetic energy begins being pumped into the lower harmonics, this process, as shown by calculations, being gradual.

The numerical solution in a wide range of parameters of the problem of the firehose instability of the initial Alfven wave with a circular polarization in the presence of noises has made it possible to come to the following conclusions. At the times of the order of $10 \div 20\gamma_0^{-1}$ there are regular non-linear oscillations that are in good agreement with the analytical solution. At greater times there occurs stochastization of the magnetic field and the transition to the quasi-stationary regime, in which case the level of the averaged turbulent magnetic field increases with growing degree of the plasma anisotropy. In the course of the turbulence evolution the magnetic field energy is pumped from the basic harmonic, set by the initial disturbance, into the long-wave part of the spectrum.

2.3. Waves on the Surface of Viscous Films

Chapter 1 dwelt upon a linear analysis of the problem of instability of the surfaces of viscous films falling along oblique planes. Let us now consider the disturbances of the stationary solution (1.40) $\hat{u}$, $\hat{v}$, $\hat{w}$, $\hat{p}$, setting no assumptions as to their smallness [74, 80, 86, 53, 91, 103]. By substituting the expressions $(u, v, w) = (\bar{u}, \bar{v}, \bar{w}) + (\hat{u}, \hat{v}, \hat{w})$, $p = \bar{p} + \hat{p}$ into the Navier–Stokes equation for an incompressible fluid we get the system of equations for the disturbances:

$$\hat{u}_x + \hat{v}_y + \hat{w}_z = 0,$$

$$\hat{u}_t + (\bar{u} + \hat{u})\,\hat{u}_x + (\bar{u}_y + \hat{u}_y)\,\hat{v} + \hat{w}\hat{u}_z = -\hat{p}_x/\varrho + v\,\Delta\hat{u},$$

$$\hat{v}_t + (\bar{u} + \hat{u})\,\hat{v}_x + \hat{v}\hat{v}_y + \hat{w}\hat{v}_z = -\hat{p}_y/\varrho + v\,\Delta\hat{v},$$

(2.36)

$$\hat{w}_t + (\bar{u} + \hat{u})\hat{w}_x + \hat{v}\hat{w}_y + \hat{w}\hat{w}_z = -\hat{p}_z/\varrho + v\,\Delta\hat{w}.$$

Let us write the boundary conditions. Let $y = h(x, z, t)$ be the equation of the free surface of the film, n — the normal to this surface, $\mathcal{H}$ — the curvature, T — the coefficient of the surface tension, P — the tensor of viscous stress. The condition of the continuity of the normal pressure on a free surface will be as follows:

$$p_n = \mathbf{n} \cdot P\langle \mathbf{n} \rangle = -\pi + T\mathscr{H}, \qquad \text{or}$$

$$(1 + h_x^2 + h_z^2)^{-1} (p_{11}h_x^2 + p_{22} + p_{33}h_z^2 - 2p_{12}h_x + 2p_{13}h_x h_z - 2p_{23}h_z)$$
$$= -\pi + T(1 + h_x^2 + h_z^2)^{-3/2} [h_{xx}(1 + h_z^2) + h_{zz}(1 + h_x^2) - 2h_x h_z h_{xz}].$$

The equality to zero of the two independent tangential stresses on a free surface affords:

$$-h_x h_z p_{11} + h_x h_z p_{33} + h_z p_{12} + (h_x^2 - h_z^2) p_{13} - h_x p_{23} = 0, \qquad (2.37)$$

$$h_x^2 p_{11} - (h_x^2 + h_z^2) p_{22} + h_z^2 p_{33} + h_x(h_x^2 + h_z^2 - 1) p_{12}$$
$$+ 2h_x h_z p_{13} + h_z(h_x^2 + h_z^2 - 1) p_{23} = 0. \qquad (2.38)$$

Formulae (2.37), (2.38) employed the following definitions:

$$p_{11} = -\bar{p} - \hat{p} + 2\varrho v(\bar{u} + \hat{u})_x, \qquad p_{22} = -\bar{p} - \hat{p} + 2\varrho v \hat{v}_y,$$
$$p_{12} = \varrho v[\hat{v}_x + (\bar{u} + \hat{u})_y], \qquad p_{23} = \varrho v(\hat{w}_y + \hat{v}_z),$$
$$p_{13} = \varrho v[(\bar{u} + \hat{u})_z + \hat{w}_x], \qquad p_{33} = -\bar{p} - \hat{p} + 2\varrho v \hat{w}_z.$$

The kinematic condition on a free surface $y = h(x, z, t)$ is the following:

$$h_t + (\bar{u} + \hat{u}) h_x + \hat{w} h_z - \hat{v} = 0. \qquad (2.39)$$

Besides, it is necessary to add the condition of no-slip on a solid surface $y = 0$:

$$\hat{u} = \hat{v} = \hat{w} = 0. \qquad (2.40)$$

Now let us go over to the dimensionless variables, marked with primes, setting

$$x = l_0 x', \qquad y = h_0 y', \qquad z = l_0 z',$$
$$\hat{u} = U_0 u', \qquad \hat{v} = (h_0 U_0 / l_0) v', \qquad \hat{w} = U_0 w', \qquad (2.41)$$
$$t = (l_0 / U_0) t', \qquad \hat{p} = \varrho g h_0 \sin \theta \cdot p', \qquad \bar{u} = U_0 \bar{u}',$$

where $U_0 = (g h_0^2 / 3v) \sin \theta$. Let us also introduce $Re = U_0 h_0 / v$ — the Reynolds number, $\mathscr{W} = T/\varrho g h_0^2$ — the Weber number, $\mu = h_0 / l_0$.

Let us write the system of equations (2.36) and the boundary conditions (2.37)–(2.40) in the dimensionless variables (2.41), omitting, for simplicity, the primes:

$$u_x + v_y + w_z = 0,$$

$$u_{yy} = \mu\, Re[u_t + (\bar{u} + u)\, u_x + v(\bar{u} + u)_y + wu_z]$$
$$+ 3\mu p_x - \mu^2(u_{xx} + u_{zz}),$$

$$3p_y = \mu v_{yy} - \mu^2\, Re[v_t + (\bar{u} + u)\, v_x + vv_y + wv_z] + \mu^3(v_{xx} + v_{zz}),$$

$$w_{yy} = \mu\, Re[w_t + (\bar{u} + u)\, w_x + vw_y + ww_z] + 3\mu p_z$$
$$- \mu^2(w_{xx} + w_{zz}),$$

$$\bar{u} = 3y(1 - y/2);$$

$$h_z(\bar{u} + u)_y - h_x w_y + \mu^2[2h_x h_z(w_z - u_x) + (h_x^2 - h_z^2)\,(u_z + w_x)$$
$$+ h_z v_x - h_x v_z] = 0 \quad \text{at} \quad y = h, \tag{2.42}$$

$$- [h_x(\bar{u} + u)_y + h_z w_y] + \mu^2[2h_x^2(u_x - v_y) + 2h_z^2(w_z - v_y)$$
$$- h_x v_x - h_z v_z + h_z(h_x^2 + h_z^2)\, u_y + h_z(h_x^2 + h_z^2)\, u_y$$
$$+ h_z(h_x^2 + h_z^2)\, w_y + 2h_x h_z(u_z + w_x)]$$
$$+ \mu^4[h_x(h_x^2 + h_z^2)\, v_x + h_z(h_x^2 + h_z^2)\, u_z] = 0 \quad \text{at} \quad y = h,$$

$$-p + (y - 1)\, ctg\, \theta + \mu[1 + \mu^2(h_x^2 + h_z^2)]^{-1}[\mu^2 h_x^2 u_x + v_y$$
$$+ \mu^2 h_z^2 w_z - h_x(\mu^2 v_x + u_y) - h_z(\mu^2 v_x + w_y) + \mu^2 h_x h_z(u_z + w_x)]$$
$$+ \mu^2 \mathscr{W}[1 + \mu^2(h_x^2 + h_z^2)]^{-3/2}\, csc\theta \cdot [h_{xx} + h_{zz} + \mu^2(h_{xx}h_z^2 + h_{zz}h_x^2)$$
$$- 2\mu^2 h_x h_z h_{xz}] = 0 \quad \text{at} \quad y = h,$$

$$h_t + uh_x + wh_z + \bar{u}h_x - v = 0 \quad \text{at} \quad y = h,$$

$$u = v = w = 0 \quad \text{at} \quad y = 0.$$

Let us consider then the long-wave approximation, setting that the parameter $\mu = h_0/l_0$ is small, i.e. $\mu \ll 1$ (here h_0 is the undisturbed thickness of the film, l_0 is the characteristic wavelength). Besides, let us consider the motions with the Reynolds numbers $Re \sim 1$ and the Weber numbers $\mathscr{W} \sim \mu^{-2} \gg 1$. The solution is sought as an expansion in the powers of the small parameter μ:

$$f = f^{(0)} + \mu f^{(1)} + \mu^2 f^{(2)} + \cdots,$$

where $f = (u, v, w, p, h)$. Substituting these expansions into the equations and boundary conditions (2.42) and equating the terms of one order of smallness with respect to μ, we find, respectively, $f^{(0)}, f^{(1)}, \ldots$. If we set that the deviations of the film surface from the undisturbed level h_0 are small, but finite, then for the quantity ζ we get the equation

$$\zeta_t + 3\zeta_x + 6\zeta\zeta_x + \alpha\zeta_{xx} - ctg\theta \cdot \zeta_{zz}$$

$$- \beta \left(\frac{\partial^2}{\partial x^2} + \frac{\partial^2}{\partial z^2} \right)^2 \zeta = 0,$$

(2.43)

where $\zeta \equiv \dfrac{h}{h_0} - 1 \sim \mu$, $\alpha = \dfrac{6}{5} Re - ctg\theta$, $\beta = \dfrac{\mathscr{W}}{\sin\theta}$. If we then neglect the non-linear term $6\zeta\zeta_x$ and assume independence from the coordinate z, then equation (2.43) is transformed into the linear equation (1.41), that has been analysed in Chapter 1.

Figure 5 presents the results of the numerical solution of the one-dimensional analogue of equation (2.43) in the interval $x = 0 \div L$ under the following conditions:

$$u(x, 0) = 0, \quad u(0, t) = A \sin \omega t, \quad u_x(0, t) = -kA \cos \omega t,$$

$$u_{xx}(L, t) = 0, \quad u_{xxx}(L, t) = 0$$

and the parameters $\theta = 90°$ (a vertical wall), $Re = 3.578$, $\mathscr{W} = 36$ [80]. The results of the numerical solution of the same equation under the periodical boundary conditions and the initial condition $u(x, 0) = A \sin (2\pi x/L)$ [80] (the parameters θ, Re, $\mathscr{W}$ are the same as in the former case) are given in Fig. 6. Use was made of the difference scheme

$$u_j^{n+1} = u_j^n - (3\tau/2h) [u_{j+1}^n - u_{j-1}^n + (u_{j+1}^n)^2 - (u_{j-1}^n)^2$$

$$- (\alpha\tau/h^2) (u_{j+1}^{n+1} - 2u_j^{n+1} + u_{j-1}^{n+1})$$

$$- (\beta\tau/h^4) (u_{j+2}^{n+1} - 4u_{j+1}^{n+1} + 6u_j^{n+1} - 4u_{j-1}^{n+1} + u_{j-2}^{n+1}),$$

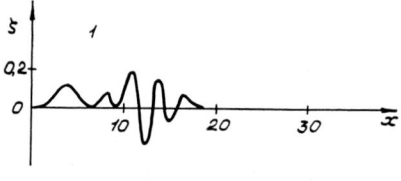

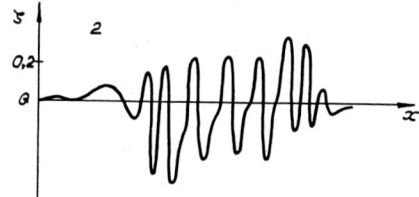

Fig. 5. Profile of a viscous film surface. $1 - t = 4$, $2 - t = 7$.

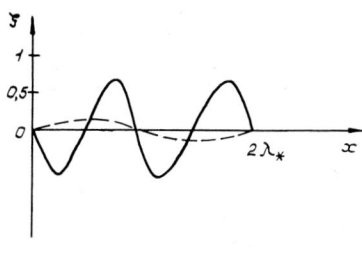

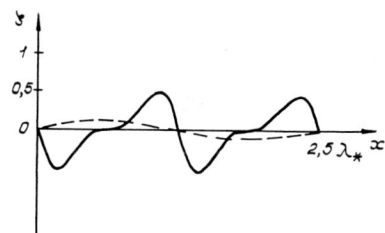

Fig. 6. Profile of a viscous film surface.

with the order of approximation $0(\tau, h^2)$. To realize this scheme, the algorithm of the five-point sweeping was used in the first problem, and the algorithm of the cyclic five-point sweeping in the second problem.

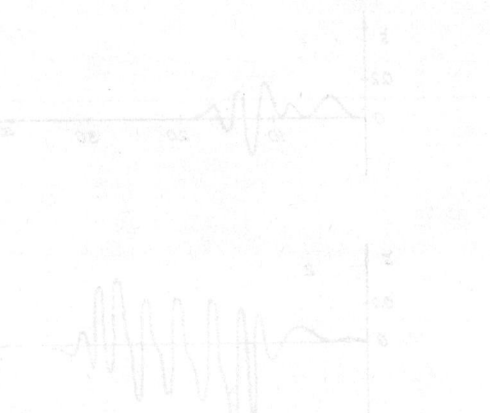

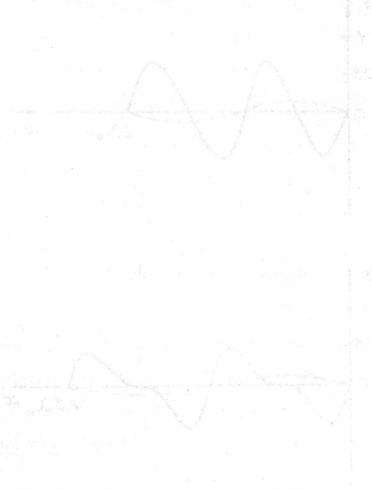

Chapter 3

Non-Linear and Shock Waves in a Rarefied Plasma (gas-dynamic models)

Plasma is a medium wherein the wave processes of various types are easily excited. Plasma dynamics, being a part of continuum mechanics, has its peculiarities distinguishing it essentially from hydro- and gas-dynamics. First, the character of the forces of interaction among charged particles (the Coulomb forces) has been definitely established and, hence, the governing mathematical model of the completely ionized plasma — the kinetic equations for ions and electrons modified by the Maxwell equations — is firmly substantiated. Second, plasma is always 'immersed' into the magnetic field, being its indispensible part, which accounts for a great variety of processes sufficiently varying in their space–time scales. Third, a very important role in plasma, especially in a rarefied plasma, belongs to the processes known as collective. Plasma is a medium where instabilities, regular and irregular intensive oscillations arise. Non-linear interactions of these oscillations and waves with one another and with the particles, which result in changes in the plasma macroscopic characteristics, are called collective processes.

The most general approach to the solution of the problem of the theory of the two-component completely ionized plasma is to make use of the kinetic equations for ions and electrons with account taken of the self-consistent electromagnetic fields

$$\frac{\partial f_{i,e}}{\partial t} + v \frac{\partial f_{i,e}}{\partial r} + (F_{i,e}/m_{i,e}) \frac{\partial f_{i,e}}{\partial v} = St_{i,e},$$

$$\text{rot } B = \mu_0 \sum_{i,e} q_{i,e} \int f_{i,e} v \, dv + \mu_0 \varepsilon_0 \frac{\partial E}{\partial t},$$

$$\text{rot } E = -\partial B/\partial t, \quad \text{div } B = 0, \tag{3.1}$$

$$\varepsilon_0 \text{ div } E = \sum_{i,e} q_{i,e} \int f_{i,e} \, dv,$$

$$F_{i,e} = q_{i,e}(E + [vB]).$$

Here $f_{i,e}$ is the function of ion or electron velocity distribution, $q_{i,e}$ is the ion or electron charge, $m_{i,e}$ is the ion or electron mass, B is the magnetic field strength, E is the electric field strength, $St_{i,e}$ is the integral of collisions, allowing for close collisions of the particles with one another.

A rarefied plasma, where the length of the free run of the particles of comparatively close (Coulomb) collisions is great as compared to the characteristic spatial scales, which essentially change the macroscopic characteristics, in the most general case can be considered using the Vlasov kinetic equations, i.e. equations (3.1) at $St_{i,e} = 0$. The most universal and widely applied recent method for solving the problems of the theory of plasma is the method of discrete modelling, or the method of particles. Plasma is described as a set of a sufficiently great number of model particles with their trajectories being the characteristics of the kinetic equations. The particle motion takes place, according to the laws of classical mechanics, in a self-consistent electromagnetic field, determined from the Maxwell equations using the charges and currents as sources. The densities of these charges and currents, in their turn, are calculated by the coordinates and velocities using some procedure. However, numerical realization of the completely kinetic model (3.1), even using the most powerful modern computers, is hampered by very serious difficulties because of too great differences in the scales of the processes. Therefore, there have been developed and are widely used some simpler but contensive and still fairly complex plasma models which can be subdivided into two groups: 'gas-dynamical' and combined (hybrid).

Gas-dynamical plasma models are formulated on the basis of the system of equations for the moments of the functions of the particle velocity distribution, and the Maxwell equations. Let us first of all consider a dense plasma wherein the length of the particle free path l is small as compared to the characteristic linear dimension L, while the characteristic duration of the processes t_p is great as compared to the time τ between the collisions. In this case the distribution functions $f_{i,e}$ for the ions and electrons can be expanded into series with respect to the small parameter l/L or τ/t_p; the Maxwell function being the zero term of the expansion for every plasma component. Subsequent to certain transformations, we get the following equations of gas-dynamic type:

$$\frac{\partial n}{\partial t} + \text{div}\,(nv_e) = 0, \qquad \frac{\partial n}{\partial t} + \text{div}\,(nv_i) = 0,$$

$$nm_e \left(\frac{\partial v_e}{\partial t} + (v_e\nabla)\,v_e\right) = -ne(E + [v_eB]) - \nabla p_e + F,$$

$$nm_i \left(\frac{\partial v_i}{\partial t} + (v_i\nabla)\,v_i\right) = ne(E + [v_iB]) - \nabla p_i - F,$$

$$\frac{3}{2}\, n\left(\frac{\partial T_e}{\partial t} + v_e\,\nabla T_e\right) + p_e\,\text{div}\,v_e = Q_e - \text{div}\,q_e, \qquad (3.2)$$

$$\frac{3}{2}\, n\left(\frac{\partial T_i}{\partial t} + v_i\,\nabla T_i\right) + p_i\,\text{div}\,v_i = Q_i - \text{div}\,q_i,$$

$$p_e = nT_e, \qquad p_i = nT_i,$$

$$\text{rot}\,B = \mu_0 en(v_i - v_e)$$

$$\frac{\partial B}{\partial t} = -\text{rot}\,E, \qquad \text{div}\,B = 0.$$

The following denotations are used in equations (3.2): n is the number of electrons or ions in a unit volume, $v_e(v_i)$ is the macroscopic velocity of electrons (ions), $T_e(T_i)$ is the temperature of electrons (ions) in energy units, $p_e(p_i)$ is the pressure of electronic (ionic) gas, B is the magnetic field strength, E is the electric field strength. The plasma is considered completely ionized, quasi-neutral $(n_i = n_e)$ and hydrogenous (electrons and protons); the tensor of viscous stresses and the displacement currents are neglected, the ratio of heat capacities under constant pressure and constant volume γ is set equal to 5/3 (one-atom gas).

In equations of motion the quantity F denotes the momentum transfer from one plasma component to another one during a second in a unit volume. In other words, it is a macroscopic force of friction between ions and electrons, which consists of two parts: the first is determined by the presence of the relative velocity $u \equiv v_e - v_i$, the second by the gradient of electron temperature (thermo-force), i.e.

$$F = F_u + F_T,$$

$$F_u = (\alpha_\| / en)\, j_\| + (\alpha_\perp / en)\, j_\perp - (\alpha / en)\, [bj],$$

$$F_T = -\beta_\| \nabla_\| T_e - \beta_\perp \nabla_\perp T_e - \beta[b\nabla T_e], \qquad (3.3)$$

$$j = -enu, \qquad b = B/|B|,$$

$$j_\| = (jb)\, b, \qquad j_\perp = [b[jb]].$$

Hereafter the vector indices $\|, \perp$ serve to denote the components of these vectors along and across the magnetic field. In the equations of energy the electron and ion heat fluxes are determined by the following formulae

$$q_e = q^u + q^T,$$

$$q^u = \bar\beta_\| u_\| + \bar\beta_\perp u_\perp + \bar\beta[bu], \qquad (3.4)$$

$$q^T = -\mathcal{H}^e_\| \nabla_\| T_e - \mathcal{H}^e_\perp \nabla_\perp T_e - \mathcal{H}^e[b\,\nabla T_e],$$

$$q_i = -\mathcal{H}^i_\| \nabla_\| T_i - \mathcal{H}^i_\perp \nabla_\perp T_i + \mathcal{H}^i[b\,\nabla T_i].$$

The quantity of heat evolved in ion and electron gases due to collisions of the particles of various types is governed by the formulae

$$Q_i = (3nm_e\nu_e/m_i)\,(T_e - T_i) \tag{3.5a}$$

$$Q_e = -(Q_i + Fu). \tag{3.5b}$$

In formulae (3.3)–(3.5) the following denotations are used: ν_e is the frequency of electron collisions with ions, ν_i is the frequency of ion collisions with ions,

$$\alpha_\| = 0.51 m_e n \nu_e, \quad \alpha_\perp = m_e n \nu_e, \quad \alpha = 1.7 \alpha_\perp \nu_e/\omega_B, \quad \beta_\| = 0.71 n,$$

$$\beta_\perp = 5.1 n(\nu_e/\omega_B)^2, \quad \beta = 1.5 n \nu_e/\omega_B, \quad \bar{\beta}_{\|,\perp} = \beta_{\|,\perp} T_e, \quad \bar{\beta} = \beta T_e,$$

$$\mathscr{H}^e_\| = 3.16 n T_e/m_e \nu_e, \quad \mathscr{H}^e_\perp = 1.47 \mathscr{H}^e_\|(\nu_e/\omega_B)^2, \quad \mathscr{H}^e = 0.79 \mathscr{H}^e_\| \nu_e/\omega_B,$$

$$\mathscr{H}^i_\| = 3.9 n T_i/m_i \nu_i, \quad \mathscr{H}^i_\perp = 0.51 \mathscr{H}^i_\|(\nu_i/\Omega_B)^2, \quad \mathscr{H}^i = 0.64 \mathscr{H}^i_\| \nu_i/\Omega_B.$$

These expressions for the coefficients α, β, $\bar{\beta}$, $\mathscr{H}^e$, $\mathscr{H}^i$ correspond to the case of the magnetized plasma ($\nu_e/\omega_B \ll 1$, $\nu_i/\Omega_B \ll 1$); general formulae are given in [36].

For the Coulomb collisions of particles with one another we have

$$\nu_e = \frac{4\pi^{1/2} \Lambda n e^4}{3 m_e^{1/2} T_e^{3/2}}, \qquad \nu_i = \frac{4\pi^{1/2} \Lambda n e^4}{3 m_i^{1/2} T_i^{3/2}} \tag{3.6}$$

where Λ is the Coulomb logarithm. The physical sense of the transport coefficients dependence on the plasma parameters is given in detail in [36]. In [42, 44] one can find numerical solutions of the axially-symmetrical one-dimensional equations of magnetohydrodynamics with the non-linear dissipative processes accounted for (ion viscosity, ion and electron heat conductivities, resistivity, energy exchange among the plasma components), characteristic for the two-tempera-ture plasma and the experiment conditions. Based on these calcula-tions, the one-dimensional theory of pinch-effect in plasma has been created [44]. In the Institute of Applied Mathematics of the USSR Academy of Sciences a team of workers headed by academician A. N. Tikhonov and academician A. A. Samarsky has developed the numerical methods, created a complex of programmes, and carried out a bulky series of calculations of the wide range of phenomena described by the single-fluid one-dimensional magnetohydrodynamic model, taking into account both heat-conductivities and the conductivities depending on plasma density and temperature [87, 95–97, 102]. On the basis of these calculations there has been discovered the phenomenon of the T-layer — a high-temperature electro-conducting gas layer, resulting from competitive interaction of non-linear processes of the Joule heating and heat conductivity. A series of calculations of the axially-symmetrical radial plasma pinch in a magnetic field is presented in [54, 88]. These numerical methods have also found their fruitful use in studying the hydrodynamical stage of the plasma focus [43].

Let us now consider the antipodal case of a rarefied plasma, when the length of the free path of plasma particles with respect to close collisions l is great as compared to the characteristic spatial scales L, which have been experimentally detected when studying cosmic and laboratory plasmas [24, 93]. In this case the effects associated with collective interaction of the magnetic field and particles become determining [93]. It appears that for a rarefied plasma one can also formulate a model of the hydrodynamical type on the basis of (1) the system of equations for the moments of the functions of particle distribution in a phase space with account taken of their distortion under the effect of small-scale waves, related to the evolution of various plasma instabilities, and (2) the Maxwell equations for electro-magnetic fields.

Within such an approach the collective effects switch on self-consistently through the 'turbulent' transport coefficients (conductivity, heat conductivity) which arise due to plasma particle interactions with the fluctuation fields of instabilities [50, 60, 73]. For them to be calculated it is necessary to know the parameters of the most probable and important for concrete problem microinstabilities (threshold of excitation, spectrum evolution, level of fluctuation saturation), which, in their turn, are determined by both complex interaction of these waves with one another and inverse effects of the plasma macroscopic characteristics on the turbulent microprocesses.

In many problems, however, such an approach proves to be 'excessively accurate'; and, hence, use is made of approximate estimates, and the evolution of the particles momentum in the process of their dissipation due to the electro-magnetic field fluctuations is approximated by the expression analogous to the macroscopic force of friction between electrons and ions, $F = m_e v_{ef}(v_e - v_i)$, where v_{ef} is the effective frequency of the electron 'collisions' with the fluctuation fields, related to collective interactions. In this case this effective frequency of collisions becomes a part of the plasma conductivity and heat conductivity. In order to account for small effects of the Coulomb collisions, we can present the effective frequency of collisions as a sum of the terms describing Coulomb and collective interactions, i.e. $v_{ef} = v_e + v_c$, where v_c depends on a concrete problem. According to the works by R. Z. Sagdeev, in a rarefied plasma shock waves with the front width much less than the length of the free path of the particles (collisionless shock waves) can be formed, which is determined by the three factors. The first factor is non-linearity resulting in an increase of the steepness of the disturbance forefront; this phenomenon also being typical of ordinary gas dynamics. The second factor is the dispersion of small-amplitudes waves, and the third factor is the collisionless energy dissipation in the wave front resulting from the development of small-scale instabilities. Both dispersion and collisionless dissipation can result in widening the disturbance front.

Competition between the non-linearity effects and dispersion results in the possibility of forming the waves of a finite amplitude, propagating in plasma with a constant velocity without changing the form. The characteristic spatial scale of such waves is of the same order as the length of dispersion δ. The stationary waves in a two-component plasma with the laws of dispersion of the type described earlier have been studied in detail (see, for instance, [93]). As an interesting example of the interaction of dispersion and non-linearity let us consider the non-linear plane waves propagating perpendicular to the magnetic field in a quasi-neutral cold plasma ($p \ll B^2/2\mu_0$), consisting of the electrons and ions of two kinds, differentiated by their masses (a three-component plasma) [26]. We shall consider the low-frequency motions with $\omega \ll eB/(m_i m_e)^{1/2}$ and, therefore, for describing the electron motions we can use the drift approximation: $v_e = [EB]/B^2$. In this case the corresponding system of equations, in the absence of any dissipation, is as follows

$$\frac{\partial n_j}{\partial t} + \frac{\partial}{\partial x}(n_j u_j) = 0, \qquad \frac{\partial n_e}{\partial t} + \frac{\partial}{\partial x}(n_e E_y/B) = 0$$

$$m_j\left(\frac{\partial u_j}{\partial t} + u_j\frac{\partial u_j}{\partial x}\right) = e(E_x + v_j B),$$

$$m_j\left(\frac{\partial v_j}{\partial t} + u_j\frac{\partial v_j}{\partial x}\right) = e(E_y - u_j B) \qquad (3.7)$$

$$\frac{\partial B}{\partial t} = -\frac{\partial E_y}{\partial x}, \qquad \frac{\partial B}{\partial x} = -\mu_0 e(n_1 v_1 + n_2 v_2 + n_e E_x/B)$$

$$n_1 + n_2 = n_e$$

where the index j denotes the kind of ions, the z-axis is directed along the magnetic field, the ions are assumed to be once ionized.

Linearizing the system of equations (3.7), we get the law of dispersion

$$k^2 = (\alpha_1 + \alpha_2/\mu)^{-1}(\omega/V_A)^2[(\mu\alpha_1 + \alpha_2)\omega_1^2 - \omega^2] \qquad (3.8)$$
$$\times[\mu(\mu\alpha_1 + \alpha_2)\omega_1^2 - (\alpha_1 + \mu\alpha_2)\omega^2]^{-1}$$

Here $\omega_j = eB_0/m_j$, $\mu = m_1/m_2$, $\alpha_j = n_{j0}/n_0$, n_{j0}, n_{j0} is the undisturbed density of the ions, n_0 is the undisturbed density of the electrons, $V_A = B_0[\mu_0 n_0(m_1\alpha_1 + m_2\alpha_2)]^{-1/2}$.

The qualitative character of the law of dispersion $\omega = \omega(k)$ is presented in Fig. 7. Hereafter the index 1 will refer to heavier ions. At low frequencies the phase velocity of small-amplitude waves is roughly constant and equals V_A. As the frequency approaches the value $\omega = \omega_* = \left(\omega_1\omega_2\dfrac{\mu\alpha_1 + \alpha_2}{\alpha_1 + \mu\alpha_2}\right)^{1/2}$, the phase velocity begins decreasing, becoming zero at $\omega = \omega_*$. At higher frequencies there is the second branch of oscillations for which $\omega(0) = (\mu\alpha_1 + \alpha_2)\omega_1 > \omega_*$ basically conditioned by the motion of the lighter type of ions. With growing frequency ($\omega \gg \omega_*$) the upper branch of the curve of dispersion has the asymptote $\omega/k = V_A(\alpha_1 + \alpha_2/\mu)^{1/2} (\alpha_1 + \mu\alpha_2)^{1/2}$. The phase velocity deviation from the value determined by the above expression becomes essential at $\omega \to eB_0/(m_i m_e)^{1/2}$ and is conditioned by the electron inertia but this case has been discussed in detail earlier.

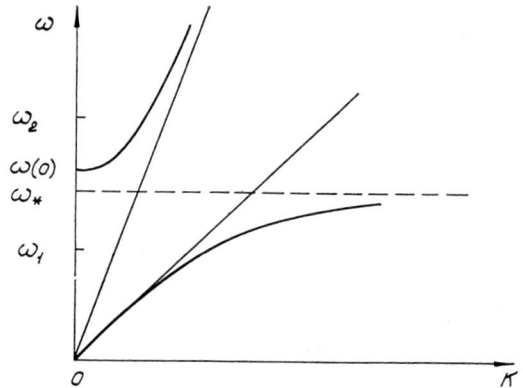

Fig. 7. Dispersion law for a three-component plasma.

At comparatively small wave velocities, it is the lower branch of the dispersion curve that 'works' and by the order of magnitude the dispersion length is

$$\delta \sim V_A/\omega_* = (c/\omega_{0e}) (\alpha_1 + \mu\alpha_2)^{1/2} (\mu\alpha_1 + \alpha_2)^{-1} (m_1/m_e)^{1/2} \gg c/\omega_{0e}.$$

If $m_1\alpha_1 \gg m_2\alpha_2$ (small concentration of the light component or a substantial difference in the ion masses), then $\delta \sim c(\varepsilon_0 m_2/n_{10}e^2\alpha_1)^{1/2}$. In another limiting case, when $m_1\alpha_1 \ll m_2\alpha_2$ (concentration of heavy ions is small), we have $\delta \sim c(\varepsilon_0 m_1/n_{20}e^2)^{1/2} (m_1/m_2)^{1/2}$. With growing wave velocity the second branch of the dispersion curve switches on and the character of the law of dispersion changes.

Let us now begin studying steady non-linear waves, which can be easily done in the system of coordinates, moving with the wave velocity U. Directing the x-axis towards the plasma motion in front of the wave, at $x \to -\infty$ we shall have an undisturbed plasma with the parameters $n_j = n_{j0}$, $n_e = n_0$, $u_j = U$, $v_j = 0$, $E_x = 0$, $B = B_0$. Through simple transformations system (3.7) is reduced to a single equation for the magnetic field strength:

$$\delta^2 u^2 \frac{d}{dx}\left\{\frac{u}{ub - \alpha_1 M}\left(1 + \frac{\mu u}{\mu\alpha_1 + \alpha_2}\frac{du}{db}\right)\frac{db}{dx}\right\} = M - ub, \qquad (3.9)$$

where $M = U/V_A$, $b = B/B_0$,

$$u = u_1/V_A = (M/2m_1\alpha_1 b)\left\{m_1\alpha_1^2 - m_2\alpha_2^2 + (m_1\alpha_1 + m_2\alpha_2)\right.$$

$$\times \left(1 + \frac{1 - b^2}{2M^2}\right)b + \left[(m_1\alpha_1^2 - m_2\alpha_2^2)^2\right.$$

$$- 2(m_1\alpha_1 + m_2\alpha_2)(m_1\alpha_1^2 + m_2\alpha_2^2)\left(1 + \frac{1 - b^2}{2M^2}\right)b \qquad (3.10)$$

$$\left.\left. + (m_1\alpha_1 + m_2\alpha_2)^2\left(1 + \frac{1 - b^2}{2M^2}\right)^2 b^2\right]^{1/2}\right\}$$

When the condition $m_1\alpha_1 \gg m_2\alpha_2$ is met, expression (3.10) for the velocity of heavy ions is essentially reduced: $u = M\left(1 + \frac{1 - b^2}{2M^2}\right)$. Then by way of integrating (3.9), we get

$$(\delta^2 u^2/2)(ub - \alpha_1 M)^{-2}\left(1 + \frac{\mu u}{\mu\alpha_1 + \alpha_2}\frac{du}{db}\right)^2\left(\frac{db}{dx}\right)^2$$

$$= -\left(1 + \frac{\mu\alpha_2}{\mu\alpha_1 + \alpha_2}\right)(b - 1) - \frac{\mu/2}{\mu\alpha_1 + \alpha_2}(u^2 - M^2)$$

$$- \frac{2\alpha_2^2 M^2}{\mu\alpha_1 + \alpha_2}\left\{(b_2 - b_1)^{-1}(b_3 - b_1)^{-1}\ln\left|\frac{b - b_1}{1 - b_1}\right|\right. \qquad (3.11)$$

$$+ (b_1 - b_2)^{-1}(b_3 - b_2)^{-1}\ln\left|\frac{b - b_2}{1 - b_2}\right|$$

$$\left. + (b_2 - b_3)^{-1}(b_1 - b_3)^{-1}\ln\left|\frac{b - b_3}{1 - b_3}\right|\right\}$$

where b_1, b_2, b_3 are the roots of the equation $b^3 - (2M^2 + 1)\,b + 2M^2\alpha_1 = 0$, and the choice of the integration constant corresponds to a solitary wave. Setting $b = b_{max}$, $db/dx = 0$ in (3.11), we can easily determine the relation between the velocity of a solitary wave M and the maximum value of the magnetic field in the wave b_{max}. This dependence for different values of the relative concentrations of the ions is shown in Fig. 8 $\left(1 - \alpha_1 = 0.9,\ \ 2 - \alpha_1 = 0.7,\ \ 3 - \alpha_1 = 0.5,\right.$

$\left. 4 - M = \dfrac{1}{2}\,(1 + b_{max})\right).$

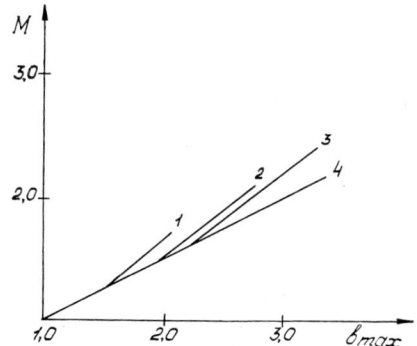

Fig. 8. Dependence of a solitary wave velocity on the magnetic field amplitude at various values of relative ion concentrations. $1 - \alpha_1 = 0.9$;

$$2 - \alpha_1 = 0.7;\ 3 - \alpha_1 = 0.5;\ 4 - M = \frac{1}{2}\,(1 + b_{max}).$$

At small amplitudes of the magnetic field the velocity of a solitary wave is

$$M = \frac{1}{2}\,(1 + b_{max}). \tag{3.12}$$

With growing amplitude b_{max} the wave velocity grows faster than by formula (3.12); the sooner this deviation begins, the less the relative concentration of lighter ions. For the case of small amplitudes $(b = 1 + \tilde{b},\ \tilde{b} \ll 1)$, by expanding equation (3.11) to the accuracy of the terms of the order of $\tilde{b}^2$, we get the solitary wave profile (symmetrical 'hump', as was the case in a two-component plasma [93]):

$$b = 1 + b_{\max} ch^{-2}\left(\sqrt{b_{\max}}\, x/2\delta\right).$$

At great wave velocities, when the dependence $M = M(b_{\max})$ deviates from formula (3.12), the profile of a solitary wave can significantly change; the effect of the upper branch of the dispersive curve can result in the appearance of a rarefied wave on the background of the compression wave.

The profiles of solitary waves for various ratios of ion masses m_1/m_2. concentrations α_1, α_2 and velocities M are determined by numerical integration of system (3.9), (3.10). Let us begin with linearizing this system in the vicinity of a singular point $u = M$, $b = 1$, corresponding to the undisturbed state of plasma in front of the wave, and studying this special point. Substituting into (3.9) the expressions $u = M + \tilde{u}$, $b = 1 + \tilde{b}$, where $\tilde{u}, \tilde{b}$ are the quantities of the first order of smallness, and assuming $\tilde{u}, \tilde{b} \sim \exp(kx)$, we find the following characteristic equation

$$k^2 = (\alpha_2/\alpha_1)(\mu\alpha_1 + \alpha_2)\frac{M^2 - 1}{\mu\alpha_2 + \alpha_1 - (\alpha_1 + \alpha_2/\mu)^{-1}M^2}. \quad (3.13)$$

Consequently, the singular point considered is a saddle and the integral curve leaves this point when the condition

$$1 < M < [(\mu\alpha_2 + \alpha_1)(\alpha_1 + \alpha_2/\mu)]^{1/2} \quad (3.14)$$

is met. Therefore, at small deviations from the undisturbed state we have

$$u = M - \frac{\mu\alpha_1 + \alpha_2 - M^2}{(\mu - 1)\alpha_1 M}C\exp(kx), \quad b = 1 + C\exp(kx) \quad (3.15)$$

where $C > 0$ is a constant, k is determined by formula (3.13). These values of the sought functions at a certain negative x are chosen as the initial conditions to solve equation (3.9) by the Runge-Kutta method of the fourth degree of accuracy. Figure 9 presents the profiles of the magnetic field in solitary waves of the considered type at different values of concentrations and velocities ($1 - \alpha_1 = \alpha_2 = 0.5$, $M = 2.2$, $\mu = 20$; $2 - \alpha_1 = 0.9$, $\alpha_2 = 0.1$, $M = 1.6$, $\mu = 20$). With decreasing concentration of light ions the linear size of the solitary wave reduces, which is in accord with the given estimate of the dispersion length; besides, the value of the 'trough' in the wave centre decreases.

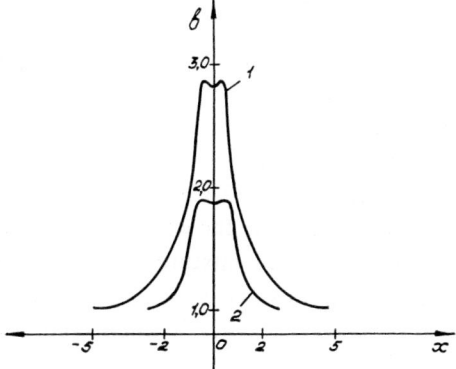

Fig. 9. Magnetic filed profiles in solitary waves. $1 - \alpha_1 = \alpha_2 = 0.5$; $M = 2.2$, $\mu = 20$; $2 - \alpha_1 = 0.9$, $\alpha_2 = 0.1$, $M = 1.6$, $\mu = 20$.

Let us now consider the problem of the limiting (critical) velocities of solitary waves, assuming for greater generality that the ions differ not only in their masses but also in their charges. As follows from the condition, generalizing (3.14) for the case when $z_1 \neq 1$, $z_2 \neq 1$, the solution describing a steady solitary wave, exists only at certain values of the wave velocity. The lower limit of the wave velocity M_- equals, naturally, the Alfven velocity (the speed of 'sound' in a cold plasma), i.e. $M_- = 1$. The upper limit of the wave velocity is

$$M_+ = [1 + \alpha_1\alpha_2 z_2^2(\mu - \zeta)^2/\mu]^{1/2}, \quad \zeta = z_1/z_2 \qquad (3.16)$$

and depends on the mass ratio, ion charges and their relative concentrations. If the wave velocity exceeds M_+, then a singular point, corresponding to the undisturbed state of the plasma in front of the wave is a centre, and there is no integral curve going out of this point (there is no wave of a steady form). The critical wave velocity M_+ equals the phase velocity of small oscillations, when $\omega \gg \omega_*$. At $M \to M_+$ the effect of the upper branch of the dispersive curve is essential, where the law of dispersion $\omega = \omega(k)$ is linear almost up to the frequencies $\omega \sim eB_0/(m_i m_e)^{1/2}$; hence, within this frequency region the dispersive effects will reveal themselves only if $\omega \to eB_0/(m_i m_e)^{1/2} \gg \omega_*$, and will be due to electron inertia. With the wave velocity approaching the upper limit M_+, an increase of the profile steepness due to non-linearity at $\omega \ll eB_0/(m_i m_e)^{1/2}$ will not be compensated for by dispersive 'smearing'.

Let us investigate the character of the critical velocity M_+ dependence on the parameters. Formula (3.16) shows that if $\mu = \zeta$, no steady wave exists, since $M_+ = 1$. As to the excitation of such waves, of no interest is a completely ionized mixture, with deuterium as a light ion component. Indeed, in this case $\mu \approx \zeta$ and the difference between the upper and lower wave velocities is small, and, hence, the wave amplitudes are small.

At small admixtures of the light or heavy component (α_1 or $\alpha_2 \ll 1$) the critical velocity of a solitary wave can be written as follows

$$M_+ \approx 1 + (\alpha z_2^2/2\mu)\,(\mu - \zeta)^2,$$

where α is a relative concentration of the heavy or light component.

At a fixed ratio of the masses μ and charges ζ, the critical value of the velocity M_+ reaches its maximum $M_+^{\max} = [1 + (\mu - \zeta)^2/4\mu\zeta]^{1/2}$ at the heavy component concentration $\alpha_1 = 1/2z_1$, i.e. under the condition that the charge is equally distributed between the components: $\alpha_1 z_1 = \alpha_2 z_2 = 1/2$. For the case of once ionized ions we have

$$M_+^{\max} = \frac{\mu + 1}{2\mu^{1/2}},$$

which demonstrates that the maximum velocity of the considered solitary waves is small provided the difference between the ion masses is small: for instance, for a mixture of hydrogen and helium $M_+^{\max} = 1.25$, and for a deuterium–tritium mixture $M_+^{max} = 1.01$. And, correspondingly, the amplitudes of the magnetic field in such waves are also small. At $\mu \gg 1$ $M_+^{max} = \frac{1}{2}\mu^{1/2}$. If $\zeta \neq 1$ and $\mu \gg \zeta$, then $M_+^{max} = \frac{1}{2}(\mu/\zeta)^{1/2}$. If ionization is complete, then $\zeta \approx \mu/2$ (provided hydrogen is chosen as the light component), therefore $M_+^{max} = 3/2^{3/2}$.

If both the ratio of the ion masses μ and the relative concentration of the heavy component α_1 are fixed, the critical velocity of a solitary wave reaches its maximum at $\zeta = [(\mu\alpha_1 z_2 + 2)/3\alpha_1]$ and $\mu\alpha_1 < 1$, where the square brackets denote the integer part, this maximum value being $M_+^{max} \approx [1 + 4(1 - \mu\alpha_1 z_2)^3/27\mu\alpha_1 z_2]^{1/2}$. This expression is obtained from (3.16) by substituting the value $\zeta = (\mu\alpha_1 z_2 + 2)/3\alpha_1$. If $\mu\alpha_1 \ll 1$ (a small admixture of the heavy component), then $M_+^{max} \approx (2/3^{3/2})(\mu\alpha_1 z_2)^{-1/2}$.

Thus, the waves of the considered type with sufficiently great amplitudes (and, hence, velocities) may arise at a single ionization and at a substantial difference in the masses of the ions comprising the plasma. Now let us estimate the ion energy in a solitary wave, denoting through $\varepsilon_x^{(1)}$ the density of the energy of heavy ions in the direction of the wave propagation ('longitudinal' energy), and through $\varepsilon_y^{(2)}$ the density of the energy of light ions in the direction perpendicular to the wave propagation ('transversal' energy). A solitary wave corresponds to the frequencies $\omega \sim \omega_* = (\omega_1\omega_2\alpha_1/\alpha_2)^{1/2}$ (when the condition $m_1\alpha_1 \gg m_2\alpha_2$ is met) and, generally speaking, $\omega_1 \ll \omega_* \ll \omega_2$. Therefore, since at $\omega \gg \omega_1, m_1\frac{du_1}{dt} \sim m_1\omega u_1 \sim eE_x$, we have

$$\varepsilon_x^{(1)} = \frac{1}{2}n_1m_1u_1^2 \sim n_1m_1(eE_x/m_1\omega_*)^2. \qquad (3.17)$$

If $\omega \ll eB_0/(m_im_e)^{1/2}$, the light ion velocity can be estimated from the drift approximation, i.e. $v_2 \sim E_x/B$, therefore

$$\varepsilon_y^{(2)} = \frac{1}{2}n_2m_2v_2^2 \sim n_2m_2(E_x/B)^2.$$

Substituting into (3.17) the expression for ω_*, we see that the energy density of the heavy ions in the longitudinal direction equals that of the light ions in the transversal direction by the order of magnitude. Comparison of the energies $\varepsilon_x^{(1)}$ and $\varepsilon_y^{(2)}$, belonging to one particle, results in $\varepsilon_y^{(2)}/\varepsilon_x^{(1)} \sim n_1/n_2$. In the case when the concentration of light ions is small as compared to that of heavy ions, the energy $\varepsilon_y^{(2)}$ of the light ions in the transversal direction is much greater than the energy $\varepsilon_x^{(1)}$ of the heavy ions in the longitudinal direction. Therefore, in the process of a solitary wave propagation in a three-component plasma, there takes place an interesting acceleration mechanism, resulting in the acceleration of light ions in the direction perpendicular to that of the wave propagation.

Small energy dissipation results in the fact that instead of the periodical and solitary waves, there will be formed shock waves with an oscillatory structure [93], relating two different plasma states — in front of the wave and behind it. If the phase velocity of small oscillations decreases with reducing wavelength, the shock wave has a sharp forefront, accompanied by an oscillatory train; if it increases with reducing wavelength, there arises a shock wave with an oscillating precursor [62, 93]. The structure of a steady shock wave in a three-component plasma is given in Fig. 10 ($\mu = 20$, $\alpha_1 = \alpha_2 = 0.5$, $M = 2.2$). The total length of the oscillation damping, which in the

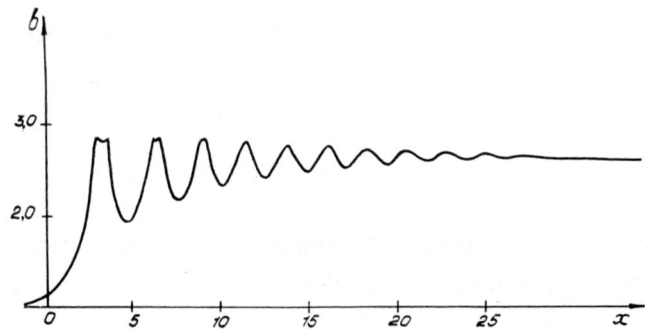

Fig. 10. Magnetic field profile in a shock wave in a three-component plasma. $\alpha_1 = \alpha_2 = 0.5$, $M = 2.2$, $\mu = 20$.

case of oscillatory structures can be referred to as the width of a shock wave front, is conditioned by the predominating dissipation mechanism.

In a highly rarefied plasma, the collisionless energy dissipation is conditioned by plasma turbulization in the wave front due to the development of the small-scale oscillations with the wavelength less than the length of dispersion of the corresponding case in question. [93]. These oscillations result in the origination of the electromagnetic field fluctuations whereon the particle dissipation takes place, which is equivalent to an essential increase of plasma resistivity (or an increase of the effective frequency of collisions ν_{ef}, provided collisions are treated as particle scattering on fluctuation fields) as compared to the case of paired collisions. Thus, the length of the plasma particle scattering on the noises is a characteristic spatial scale instead of the length of the free path in the case of paired collisions.

Collisionless shock waves have been widely studied on laboratory devices with high-temperature rarefied plasma. A bulky series of thorough experimental investigations of the shock wave structure in a rarefied plasma ($10^6 \lesssim n_0 \lesssim 10^9 \text{m}^{-3}$) in the presence of a magnetic field was carried out by Nesterikhin and Kurtmullaev with co-workers [24, 46, 49, 58, 68], Paul et al. [85] and Robson et al. [90]. In all the experiments the shock waves were excited by a 'magnetic piston' under the effect of an external pulse magnetic field. Collisionless dissipation in such waves can be associated with the development of micro-oscillations due to instability in the wave front of the electron current, the density of which exceeds a certain critical value; various types of such instabilities are considered in [2, 93, 94].

If the directed velocity $v_d = (\mu_0 en)^{-1} |\text{rot } \boldsymbol{B}|$ of electrons with respect to ions (the drift velocity of electrons) exceeds their thermal velocity $v_{T_e} = (T_e/m_e)^{1/2}$, there arises a beam instability (the Buneman instability), which is swinging of the longitudinal electrostatic plasma oscillations with the increment $\gamma_B \sim \omega_{0i} = (n_0 e^2/\varepsilon_0 m_i)^{1/2}$. The momentum loss of the electrons at their scattering on the electromagnetic field fluctuations results, as has been shown, in the collision frequency ν_c and in the appearance of the plasma resistivity σ_{ef}^{-1} due to the processes of a collective character. Such a resistivity is referred to as anomalous.

As soon as there is a resistivity, there is the Joule effect. In the absence of the Coulomb (paired) collisions this turbulent heating is not equal for the ion and electron components. As shown in [50], the Joule effect increases the temperature of electrons (if temperature is understood as a mean chaotic energy of the particles) quicker than that of ions, and there exists a universal estimate independent of the type of instability: $dT_e/dT_i \sim v_d(\omega/k)^{-1}$, where ω and k are the characteristic values of the instability. In the case of the Buneman instability we have $v_d \gtrsim v_{T_e}$, $\omega \sim \omega_{0i}$, $k \sim \omega_{0e}/v_d$, therefore, $dT_e/dT_i \sim \omega_{0e}/\omega_{0i} = (m_i/m_e)^{1/2} \gg 1$.

With growing electron temperature the condition of the Buneman instability excitation $(v_d > v_{T_e})$ may be violated, but in the non-iso-thermic plasma with $T_e \gg T_i$, provided the condition $v_d > (T_e/m_i)^{1/2} \equiv c_s$ is met, the ion-sound oscillations with the frequency $\omega \lesssim \omega_{0i}$ and the increment $\gamma_s \lesssim \omega_{0i}v_d/v_{Te}$ are excited. The electron scattering on these oscillations also contributes into the collective frequency of collisions v_c and the anomalous plasma resistivity. The ion-sound instability is self-supporting, since for it $dT_e/dT_i \sim v_d/c_s > 1$.

In the presence of a magnetic field the character of instability depends on the relation ω_{0e}/ω_B. If $\omega_{0e} \gg \omega_B$, then the magnetic field effect on the beam and ion-sound instabilities can be neglected. An important role in the creation of the anomalous resistivity can belong to the class of instabilities concerning the electrostatic disturbances with $k_{\parallel}^2 \ll k_{\perp}^2$ in the plasma with a current across the magnetic field. It is the so-called modified Buneman instability with the frequency $\Omega_B < \omega \ll \omega_B$, the increment $\gamma_{Bm} \approx (\omega_B\Omega_B)^{1/2} \ll \gamma_B, \gamma_s$ and a very low threshold of excitation $(v_d \sim v_{T_i})$. This instability arises at small currents, when more powerful instabilities (beam and ion-sound) are not excited, for instance, in the case of shock waves propagating at an angle to the undisturbed magnetic field [50]; besides, it can occur in the plasma with a high ion temperature $(T_i \sim T_e)$.

To investigate the shock waves in a rarefied plasma, let us present the effective frequency of collisions as the addends, describing the

Coulomb and collective interactions related to the development of the beam, ion-sound and modified Buneman instabilities: $\nu_{ef} = \nu_e + \nu_B + \nu_s + \nu_{Bm}$. According to [104], for the collision frequency ν_B, which is conditioned by the beam instability, we shall employ the interpolation formula

$$\nu_B = (v_d^2 - v_{T_e}^2)^{1/2} \left((m_e/m_i)^{1/3} v_d^2 + v_{T_e}^2\right)^{-1/2} \omega_{0i}, \qquad v_d \geq v_{T_e};$$

$$\nu_B = 0, \qquad v_d < v_{T_e}, \tag{3.18}$$

where $v_d = (\mu_0 e n)^{-1} |\mathrm{rot}\, \boldsymbol{B}|$, $v_{T_e} = (T_e/m_e)^{1/2}$. Formula (3.18) accounts for the threshold of excitation $v_d > v_{T_e}$; at $v_d \gg v_{T_e}$ $\nu_B \approx (m_i/m_e)^{1/6} \omega_{oi}$. Making use of the results given in [47, 50], let us write the collision frequency ν_s, associated with the ion-sound instability, as

$$\nu_s = \nu_0 (1 - K_1 T_i/T_e) \left(1 - K_2 \frac{(T_i/T_e)^{3/2}}{v_d/c_s}\right) \tag{3.19}$$

when the conditions $T_e \geq K_1 T_i$, $v_d \geq K_2 c_s (T_i/T_e)^{3/2}$ are met. When these conditions are violated, the ν_s value is set to equal zero. In formula (3.19) $K_1 = 5 \div 7$, $K_2 \approx 75$ [46, 47], $\nu_0 = 10^{-2} \dfrac{T_e}{T_i} \dfrac{v_d}{c_s} \omega_{0i}$. In line with [50], the collision frequency, conditioned by the development of the modified Buneman instability, is

$$\nu_{Bm} \approx (v_d/v_{Te}) \omega_B. \tag{3.20}$$

From now on, in numerical calculations of unsteady shock waves the quantity ν_{ef}, contained in conductivity and electron heat conductivity, will be determined by formulae (3.18)–(3.20).

Let us consider one-dimensional plane waves, assuming that the direction of the wave propagation coincides with the x-axis, all the functions depend only on the spatial coordinate x and the time t, the undisturbed magnetic field $\boldsymbol{B}_0$ lies in the plane x, z and is at the angle θ to the plane of the wave front. Let us also introduce the mean velocity

$u = (m_i v_i + m_e v_e)/(m_i + m_e) = \{u, v, w\}$, in which case

$$v_i = \{u_i, v_i, w_i\} = u + \frac{m_e}{m_i + m_e} (\mu_0 en)^{-1} \text{ rot } \boldsymbol{B},$$

$$v_e = \{u_e, v_e, w_e\} = u - \frac{m_i}{m_i + m_e} (\mu_0 en)^{-1} \text{ rot } \boldsymbol{B}.$$

Due to plasma quasi-neutrality we have $u_i = u_e = u$. Adding the equations of ion and electron motions results in

$$(m_i + m_e) n \left(\frac{\partial}{\partial t} + u \frac{\partial}{\partial x} \right) u = -\nabla(p_i + p_e) + \mu_0^{-1}[\text{rot } \boldsymbol{B} \cdot \boldsymbol{B}].$$

Let us express the components of the electric field through the equations of motion for electrons and substitute them into the equation of induction $\partial \boldsymbol{B}/\partial t = -\text{rot } \boldsymbol{E}$. After a number of transformations we get the following system of equations:

$$\frac{dn}{dt} = -n \frac{\partial u}{\partial x},$$

$$(m_i + m_e) n \frac{du}{dt} = - \frac{\partial}{\partial x} \left(nT_i + nT_e + \frac{B_y^2 + B_z^2}{2\mu_0} \right),$$

$$(m_i + m_e) n \frac{dv}{dt} = \mu_0^{-1} B_x \frac{\partial B_y}{\partial x},$$

$$(m_i + m_e) n \frac{dw}{dt} = \mu_0^{-1} B_x \frac{\partial B_z}{\partial x}, \qquad (3.21)$$

$$\frac{dB_y}{dt} = B_x \frac{\partial v}{\partial x} - B_y \frac{\partial u}{\partial x} - e^{-1} \frac{\partial}{\partial x}$$

$$\times \left\{ m_e \frac{d}{dt} \left(w - \frac{m_i}{(m_i + m_e) \mu_0 en} \frac{\partial B_y}{\partial x} \right) - \frac{m_i B_x}{(m_i + m_e) \mu_0 n} \frac{\partial B_z}{\partial x} - \frac{F_z}{n} \right\},$$

$$\frac{dB_z}{dt} = B_x \frac{\partial w}{\partial x} - B_z \frac{\partial u}{\partial x} + e^{-1} \frac{\partial}{\partial x}$$

$$\times \left\{ m_e \frac{d}{dt} \left(v + \frac{m_i}{(m_i + m_e) \mu_0 en} \frac{\partial B_z}{\partial x} \right) - \frac{m_i B_x}{(m_i + m_e) \mu_0 n} \frac{\partial B_y}{\partial x} - \frac{F_y}{n} \right\},$$

$$\frac{3}{2} n \frac{dT_e}{dt} = -nT_e \frac{\partial u}{\partial x} - \frac{\partial q_e}{\partial x} + Q_e,$$

$$\frac{3}{2} n \frac{dT_i}{dt} = -nT_i \frac{\partial u}{\partial x} - \frac{\partial q_i}{\partial x} + Q_i, \qquad (3.21)$$

$$\frac{d}{dt} \equiv \frac{\partial}{\partial t} + u \frac{\partial}{\partial x}.$$

The components of the friction force are equal to:

$$F_y = B^{-2} \left\{ (en\mu_0)^{-1} \left[\alpha_\| B_y \left(B_z \frac{\partial B_y}{\partial x} - B_y \frac{\partial B_z}{\partial x} \right) \right. \right.$$

$$- \alpha_\perp \left((B_x^2 + B_z^2) \frac{\partial B_z}{\partial x} + B_y B_z \frac{\partial B_y}{\partial x} \right) + \alpha B B_x \frac{\partial B_y}{\partial x} \right]$$

$$\left. - ((\beta_\| - \beta_\perp) B_x B_y + \beta B B_z) \frac{\partial T_e}{\partial x} \right\},$$

$$F_z = B^{-2} \left\{ (en\mu_0)^{-1} \left[\alpha_\| B_z \left(B_z \frac{\partial B_y}{\partial x} - B_y \frac{\partial B_z}{\partial x} \right) \right. \right.$$

$$+ \alpha_\perp \left((B_x^2 + B_y^2) \frac{\partial B_y}{\partial x} + B_y B_z \frac{\partial B_z}{\partial x} \right) + \alpha B B_x \frac{\partial B_z}{\partial x} \right]$$

$$\left. - ((\beta_\| - \beta_\perp) B_x B_z - \beta B B_y) \frac{\partial T_e}{\partial x} \right\},$$

$$B^2 \equiv B_x^2 + B_y^2 + B_z^2, \qquad B = (B_x^2 + B_y^2 + B_z^2)^{1/2}.$$

The x-components of the heat fluxes for electrons and ions are governed by the expressions:

$$q_e = B^{-2} \left\{ (en\mu_0)^{-1} \left[(\bar{\beta}_\| - \bar{\beta}_\perp) \left(B_x B_y \frac{\partial B_z}{\partial x} - B_x B_z \frac{\partial B_y}{\partial x} \right) \right. \right.$$

$$\left. - \bar{\beta} B \left(B_y \frac{\partial B_y}{\partial x} + B_z \frac{\partial B_z}{\partial x} \right) \right] - (\mathcal{H}_\|^e B_x^2 + \mathcal{H}_\perp^e (B_y^2 + B_z^2)) \frac{\partial T_e}{\partial x} \right\},$$

$$q_i = -B^{-2} (\mathcal{H}_\|^i B_x^2 + \mathcal{H}_\perp^i (B_y^2 + B_z^2)) \frac{\partial T_i}{\partial x}$$

These expressions include the coefficients α, β, $\bar{\beta}$, $\mathcal{H}^e$, with the Coulomb frequency of collisions ν_e replaced by the effective frequency of collisions ν_{ef}. Heat evolutions in the electron and ion gases are

$$Q_i = (3nm_e\nu_e/m_i)(T_e - T_i),$$

$$Q_e = -Q_i - (en\mu_0)^{-1}\left(F_y\frac{\partial B_z}{\partial x} - F_z\frac{\partial B_y}{\partial x}\right).$$

In the one-dimensional case considered, it would be convenient to go over to the Lagrangian coordinates q, t', wherein the requirements for such transition derivatives with respect to the Euler coordinates are written according to the formulae

$$\frac{\partial}{\partial t} + u\frac{\partial}{\partial x} = \frac{\partial}{\partial t'}, \qquad \frac{\partial}{\partial x} = V^{-1}\frac{\partial}{\partial q},$$

where $V = n_0/n$ is the dimensionless specific volume, n_0 is a certain scale of density. The system of equations (3.21) in the Lagrangian coordinates assumes the following form (the prime at the time variable is omitted):

$$\frac{\partial u}{\partial t} = -(m_i + m_e)^{-1}\frac{\partial}{\partial q}\left(\frac{T_i + T_e}{V} + \frac{B_y^2 + B_z^2}{2\mu_0 n_0}\right),$$

$$\frac{\partial v}{\partial t} = ((m_i + m_e)\mu_0 n_0)^{-1} B_x\frac{\partial B_y}{\partial q},$$

$$\frac{\partial w}{\partial t} = ((m_i + m_e)\mu_0 n_0)^{-1} B_x\frac{\partial B_z}{\partial q}, \qquad (3.22)$$

$$\frac{\partial x}{\partial t} = u, \qquad \frac{\partial V}{\partial t} = \frac{\partial u}{\partial q},$$

$$\frac{\partial}{\partial t}\left((VB_y - \frac{m_i m_e}{(m_i + m_e)\mu_0 n_0 e^2}\frac{\partial^2 B_y}{\partial q^2}\right) = \frac{\partial}{\partial q}$$

$$\times\left(B_x v + \frac{(m_i - m_e)B_x}{(m_i + m_e)e\mu_0 n_0}\frac{\partial B_z}{\partial q} + \frac{VF_z}{n_0 e}\right),$$

$$\frac{\partial}{\partial t}\left(VB_z - \frac{m_i m_e}{(m_i + m_e)\,\mu_0 n_0 e^2}\,\frac{\partial^2 B_z}{\partial q^2}\right) = \frac{\partial}{\partial q}$$

$$\times \left(B_x w - \frac{(m_i - m_e)\,B_x}{(m_i + m_e)\,e\mu_0 n_0}\,\frac{\partial B_y}{\partial q} - \frac{VF_y}{n_0 e}\right),$$

$$\frac{3}{2}\,n_0\,\frac{\partial T_e}{\partial t} = -(n_0 T_e/V)\,\frac{\partial u}{\partial q} - \frac{\partial q_e}{\partial q} + VQ_e,$$

$$\frac{3}{2}\,n_0\,\frac{\partial T_i}{\partial t} = -(n_0 T_i/V)\,\frac{\partial u}{\partial q} - \frac{\partial q_i}{\partial q} + VQ_i.$$

Now let us write in the Lagrangian coordinates the necessary for further work components of the friction force, heat fluxes and the expressions for the density of heat sources:

$$F_y = B^{-2}\left\{(en_0\mu_0)^{-1}\left[((\alpha_\| - \alpha_\perp)\,B_y B_z + \alpha BB_x)\,\frac{\partial B_y}{\partial q}\right.\right.$$

$$\left.\left. - (\alpha_\| B_y^2 + \alpha_\perp(B_x^2 + B_z^2))\,\frac{\partial B_z}{\partial q}\right] - V^{-1}((\beta_\| - \beta_\perp)\,B_x B_y + \beta BB_z)\,\frac{\partial T_e}{\partial q}\right\},$$

$$F_z = B^{-2}\left\{(en_0\mu_0)^{-1}\left[(\alpha_\| B_z^2 + \alpha_\perp(B_x^2 + B_y^2))\,\frac{\partial B_y}{\partial q}\right.\right.$$

$$\left. - ((\alpha_\| - \alpha_\perp)\,B_y B_z - \alpha BB_x)\,\frac{\partial B_z}{\partial q}\right]$$

$$\left. - V^{-1}((\beta_\| - \beta_\perp)\,B_x B_z - \beta BB_y)\,\frac{\partial T_e}{\partial q}\right\},$$

$$q_e = B^{-2}\left\{(en_0\mu_0)^{-1}\left[((\bar\beta_\| - \bar\beta_\perp)\,B_x B_y - \bar\beta BB_z)\,\frac{\partial B_z}{\partial q}\right.\right.$$

$$\left. - ((\bar\beta_\| - \bar\beta_\perp)\,B_x B_z + \bar\beta BB_y)\,\frac{\partial B_y}{\partial q}\right]$$

$$\left. - V^{-1}(\mathscr{H}_\|^e B_x^2 + \mathscr{H}_\perp^e(B_y^2 + B_z^2))\,\frac{\partial T_e}{\partial q}\right\},$$

$$q_i = -B^{-2}V^{-1}(\mathscr{H}_\|^i B_x^2 + \mathscr{H}_\perp^i(B_y^2 + B_z^2))\,\frac{\partial T_i}{\partial q},$$

$$Q_i = (3n_0 m_e v_e / m_i V)(T_e - T_i),$$

$$Q_e = -Q_i - (en_0 \mu_0)^{-1} \left(F_y \frac{\partial B_z}{\partial q} - F_z \frac{\partial B_y}{\partial q} \right).$$

Now we shall supplement the system of equations (3.22) with the initial and boundary conditions. Let at the moment $t = 0$ a homogeneous plasma with the density n_0 occupy a half-space $x \geq 0$, the magnetic field B_0 lie in the plane x, z at an angle θ to the z-axis. The initial conditions in this case will be written as

$$u = v = w = 0, \quad V = 1, \quad B_y = 0, \quad B_z = B_0 \cos \theta,$$

$$T_e = T_e^0, \quad T_i = T_i^0. \tag{3.23}$$

Then we shall consider the unsteady waves excited in plasma by a magnetic piston, and, therefore, on the moving boundary plasma-vacuum $q = 0$ let us set a changing in time magnetic field

$$B_y(0, t) = 0, \quad B_z(0, t) = B_0[\cos \theta + A(1 - e^{-\omega t})]. \tag{3.24}$$

Besides, let us assume the absence of heat through this boundary

$$\frac{\partial T_e(0, t)}{\partial q} = \frac{\partial T_i(0, t)}{\partial q} = 0. \tag{3.25}$$

Finally, on the right boundary of the calculation domain $q = q_{max}$ we shall set the conditions of the undisturbed plasma, i.e. $f(q_{max}, t) = f(q_{max}, 0)$.

The system of equations (3.22) with conditions (3.23)–(3.25) is a mathematical model for studying plane unsteady shock waves propagating in plasma at an arbitrary angle to the undisturbed magnetic field, with dispersion, conductivity and heat conductivity taken into account.

Now let us consider an axially symmetrical problem with all the required functions dependent on the radius r and the time t. The magnetic field has only the z-component, directed along the axis of the cylindrical system of coordinates, $B = \{0, 0, B\}$. Due to quasi-neutrality, the radial velocities of ions and electrons are equal. Introducing, as in the plane case, the mean velocity and excluding the electric field, we get the following equations in the Euler coordinates:

$$\frac{dn}{dt} = -n\left(\frac{\partial u}{\partial r} + \frac{u}{r}\right),$$

$$(m_i + m_e)\,n\,\frac{du}{dt} = -\frac{\partial}{\partial r}(nT_i + nT_e + B^2/2\mu_0)$$

$$+ \frac{m_i m_e}{\mu_0 n e^2 (m_i + m_e)\,r}\left(\frac{\partial B}{\partial r}\right)^2,$$

$$\frac{dB}{dt} = -\left(\frac{\partial u}{\partial r} + \frac{u}{r}\right)B + \frac{m_i m_e}{\mu_0(m_i + m_e)\,e^2}\left(\frac{\partial}{\partial r} + \frac{1}{r}\right)$$

$$\times \left(\frac{d}{dt} + \frac{u}{r}\right)\left(\frac{1}{n}\frac{\partial B}{\partial r}\right) - e^{-1}\left(\frac{\partial}{\partial r} + \frac{1}{r}\right)(F_\varphi/n), \qquad (3.26)$$

$$\frac{3}{2}\,n\,\frac{dT_e}{dt} = -nT_e\left(\frac{\partial u}{\partial r} + \frac{u}{r}\right) - \left(\frac{\partial}{\partial r} + \frac{1}{r}\right)q_e + Q_e,$$

$$\frac{3}{2}\,n\,\frac{dT_i}{dt} = -nT_i\left(\frac{\partial u}{\partial r} + \frac{u}{r}\right) - \left(\frac{\partial}{\partial r} + \frac{1}{r}\right)q_i + Q_i,$$

$$\frac{d}{dt} = \frac{\partial}{\partial t} + u\frac{\partial}{\partial r}.$$

The expressions for the φ-component of the friction force F_φ, for the radial components of heat fluxes $q_{e,i}$ and for the heat evolutions $Q_{e,i}$ are obtained through the formulae

$$F_\varphi = -\alpha_\perp (en\mu_0)^{-1}\frac{\partial B}{\partial r} - \beta\frac{\partial T_e}{\partial r},$$

$$q_e = -\bar\beta(en\mu_0)^{-1}\frac{\partial B}{\partial r} - \mathscr{H}_\perp^e\frac{\partial T_e}{\partial r},$$

$$q_i = -\mathscr{H}_\perp^i\frac{\partial T_i}{\partial r},$$

$$Q_i = (3nm_e\nu_e/m_i)\,(T_e - T_i), \qquad Q_e = -Q_i - (en\mu_0)^{-1}\,F_\varphi\frac{\partial B}{\partial r}.$$

In the Lagrangian coordinates q, t' $\left(\dfrac{\partial}{\partial t} + u\dfrac{\partial}{\partial r} = \dfrac{\partial}{\partial t'}, \dfrac{\partial}{\partial z} = \dfrac{r}{V}\dfrac{\partial}{\partial q} \right.$,

$V = \dfrac{n_0}{n}\Big)$ the system of equations (3.26) assumes the following form (the prime at the variable t' is omitted for simplicity):

$$(m_i + m_e)\frac{\partial u}{\partial t} = -r\frac{\partial}{\partial q}\left(\frac{T_i + T_e}{V} + \frac{B^2}{2\mu_0 n_0}\right)$$

$$+ \frac{m_i m_e r}{\mu_0 n_0^2 e^2 (m_i + m_e)}\left(\frac{\partial B}{\partial q}\right)^2,$$

$$\frac{\partial r}{\partial t} = u, \qquad \frac{\partial V}{\partial t} = \frac{\partial}{\partial q}(ru),$$

$$\frac{\partial}{\partial t}\left[VB - \frac{m_i m_e}{\mu_0(m_i + m_e)\, n_0 e^2}\left(r\frac{\partial}{\partial q}\, r\frac{\partial B}{\partial q} + V\frac{\partial B}{\partial q} \right) \right]$$

$$= \frac{m_i m_e}{\mu_0(m_i + m_e)\, n_0 e^2}\frac{\partial}{\partial q}\left(ru\frac{\partial B}{\partial q} \right) - n_0^{-1}e^{-1}\left(r\frac{\partial}{\partial q} + \frac{V}{r} \right)(F_\varphi V),$$

$$\frac{3}{2}\frac{\partial T_e}{\partial t} = -\frac{T_e}{V}\frac{\partial}{\partial q}(ru) - \frac{1}{n_0}\frac{\partial}{\partial q}(rq_e) + \frac{V}{n_0}Q_e, \qquad (3.27)$$

$$\frac{3}{2}\frac{\partial T_i}{\partial t} = -\frac{T_i}{V}\frac{\partial}{\partial q}(ru) - \frac{1}{n_0}\frac{\partial}{\partial q}(rq_i) + \frac{V}{n_0}Q_i,$$

$$F_\varphi = -r\left(\alpha_\perp (en_0\mu_0)^{-1}\frac{\partial B}{\partial q} + \beta V^{-1}\frac{\partial T_e}{\partial q} \right),$$

$$q_e = -r\left(\bar\beta (en_0\mu_0)^{-1}\frac{\partial B}{\partial q} + \mathcal{H}^e_\perp V^{-1}\frac{\partial T_e}{\partial q} \right), \qquad q_i = -r\mathcal{H}^i_\perp V^{-1}\frac{\partial T_i}{\partial q},$$

$$Q_i = (3n_0 m_e \nu_e/m_i V)(T_e - T_i), \qquad Q_e = -Q_i - r(en_0\mu_0)^{-1}F_\varphi\frac{\partial B}{\partial q}.$$

Let at the initial moment of time a resting plasma fill in the cylindrical chamber $0 \le r \le r_0$ with metallic walls; the magnetic field be homogeneous throughout and equal to B_0. If we are interested in the cylindrical waves excited in the plasma by a magnetic piston and converging to the system axis, then the initial and boundary conditions differ but slightly from conditions (3.23)–(3.25):

$$r(q, 0) = r^0(q), \quad u = 0, \quad V = 1, \quad B = B_0,$$

$$T_i = T_i^0, \quad T_e = T_e^0, \tag{3.28}$$

$$u(0, t) = \frac{\partial B(0, t)}{\partial r} = \frac{\partial T_{i,e}(0, t)}{\partial r} = \frac{\partial V(0, t)}{\partial r} = 0, \tag{3.29}$$

$$B(q_{max}, t) = B_0[1 - A(1 + e^{-\omega t})], \quad \frac{\partial T_{e,i}(q_{max}, t)}{\partial r} = 0. \tag{3.30}$$

If we are interested in the problem of the expansion of the plasmoid (in the one-dimensional case of a cylinder) over a less dense background, the initial conditions can be written as follows:

$$r(q, 0) = r^0(q), \quad u(q, 0) = 0, \quad B(q, 0) = B_0, \quad T_i(q, 0) = T_i^0, \tag{3.31}$$

$$V(q, 0) = \left[1 + (A_\varrho - 1) \left(1 + \exp\frac{\alpha_\varrho(r - r_1)}{r_0} \right)^{-1} \right]^{-1},$$

$$T_e(q, 0)/T_e^0 = 1 + (A_T - 1) \left(1 + \exp\frac{\alpha_T(r - r_1)}{r_0} \right)^{-1}.$$

The parameters A_ϱ, A_T show an increment of the density and electron temperature in the plasmoid compared to those in the plasma background.

The boundary conditions on the chamber axis remain the same, i.e. (3.29). The boundary conditions on the plasma–metal interface $r = r_0$ will be

$$u(r_0, t) = 0, \quad \frac{\partial T_{e,i}(r_0, t)}{\partial r} = 0, \quad \frac{\partial B(r_0, t)}{\partial r} = 0. \tag{3.32}$$

The latter condition results from the supposition on the infinite conductivity of the wall and from the generalized Ohm law. As far as, under the supposition, the wall conductivity $\sigma = \infty$, the tangential component of the electric field, in this case E_φ, is zero. In the plasma in the immediate vicinity with the wall we have $E_\varphi = 0$, since the tangential component of the electric field on the interface of the two media is continuous. The generalized Ohm law, with no account taken of viscosity and inertia terms, is as follows:

$$E_\varphi = uB + F_\varphi/en. \tag{3.33}$$

In the plasma layer in the immediate vicinity of the wall, at $r = r_0$ we have $u = 0$, $E_\varphi = 0$, and from condition (3.33) we get $F_\varphi = -\alpha_\perp (en\mu_0)^{-1} \dfrac{\partial B}{\partial r} = 0$, wherefrom we obtain condition (3.32) for the derivative of the magnetic field. The system of equations (3.27) either with conditions (3.28)–(3.30) or (3.29), (3.31), (3.32) is a mathematical model for studying unsteady cylindrical waves in the plasma immersed in a magnetic field.

For numerically solving the unsteady problem (3.22) it is convenient to use the scheme which is a generalized version of the scheme on the staggered meshes [92] used in general hydrodynamics. The values of the Euler coordinates and velocity components are calculated at the 'semi-integer' points $q_{j-\frac{1}{2}} = \left(j - \dfrac{1}{2}\right) h$, of all the rest functions — at the integer points $q_j = jh$, where h is the step of integration along the Lagrangian coordinate. Let us introduce the following denotations:

$$(x, u, v, w)_{j-\frac{1}{2}} = (x, u, v, w),$$

$$(V, B_y, B_z, T, T1)_j = (V, B_y, B_z, T, T1)$$

where $T = T_e$ is the electron temperature, $T1 = T_i$ is the ion temperature;

$$\hat{f} = f^{n+1}, \quad f = f^n, \quad f_t = (\hat{f} - f)/\tau,$$

$$f_q = (f_{j+1} - f_j)/h, \quad f_{\bar{q}} = (f_j - f_{j-1})/h,$$

$$f_{\tilde{q}} = (f_{j+\frac{1}{2}} - f_{j-\frac{1}{2}})/h, \quad f_{\mathring{q}} = (f_{j+1} - f_{j-1})/2h,$$

$$\bar{f} = \frac{1}{2}(f_j + f_{j-1}).$$

In these denotations the difference scheme is as follows:

$$u_t = -(m_i + m_e)^{-1} \left[(T + T1) V^{-1} + (2\mu_0 n_0)^{-1} (B_y^2 + B_z^2) \right]_{\tilde q},$$

$$x_t = \hat u, \quad \hat V = \hat x_{\tilde q}, \quad v_t = \left((m_i + m_e) \mu_0 n_0 \right)^{-1} B_0 B_{y\tilde q} \sin \theta,$$

$$w_t = \left((m_i + m_e) \mu_0 n_0 \right)^{-1} B_0 B_{z\tilde q} \sin \theta, \quad \left(V B_y - m_e (\mu_0 n_0 e^2)^{-1} B_{y q \tilde q} \right)_t$$

$$= \hat v_{\tilde q} B_0 \sin \theta + (e\mu_0 n_0)^{-1} B_0 B_{z q \tilde q} \sin \theta + e^{-1} F_{z\tilde q},$$

$$\left(V B_z - m_e (\mu_0 n_0 e^2)^{-1} B_{z q \tilde q} \right)_t = \hat w_{\tilde q} B_0 \sin \theta \qquad (3.34)$$

$$- (e\mu_0 n_0)^{-1} B_0 B_{y q \tilde q} \sin \theta - e^{-1} F_{y\tilde q}, \quad \frac{3}{2} T_t = -\hat T \hat V^{-1} \hat u_{\tilde q}$$

$$- n_0^{-1} \hat q_{e\tilde q} + n_0^{-1} \hat V \hat Q_e, \quad \frac{3}{2} T1_t = -\hat{T1} \hat V^{-1} \hat u_{\tilde q} - n_0^{-1} \hat q_{i\tilde q} + n_0^{-1} \hat V \hat Q_i.$$

Formulation of the initial (3.23) and boundary (3.24), (3.25) conditions is far from being difficult. In order to complete the difference formulation of the unsteady problem, we choose the right-hand boundary of the calculation interval $q_{max} = Jh$ at the point where the conditions of the undisturbed flow can be considered fulfilled during the time period of interest. Let us write the initial and boundary conditions:

$$(u, v, w)^0_{j-\frac{1}{2}} = 0, \quad x^0_{j-\frac{1}{2}} = \left(j - \frac{1}{2} \right) h,$$

$$B^0_{yj} = 0, \quad B^0_{zj} = B_0 \cos \theta, \quad V^0_j = 1, \qquad (3.35)$$

$$T^0_j = \tilde T_0, \quad T1^0_j = \tilde{T1}_0, \quad j = 1, 2, \ldots, J;$$

$$B^n_{y0} = 0, \quad B^n_{z0} = B_0 [\cos \theta + A(1 - e^{-\omega nt})], \qquad (3.36)$$

$$T^n_0 = T^n_1, \quad T1^n_0 = T1^n_1, \quad n = 0, 1, \ldots, N;$$

$$B^n_{yJ} = 0, \quad B^n_{zJ} = B_0 \cos \theta, \quad V^n_J = 1,$$

$$(u, v, w)^n_{J-\frac{1}{2}} = 0, \quad T^n_J = \tilde T_0, \quad T1^n_J = \tilde{T1}_0, \qquad (3.37)$$

$$n = 0, 1, \ldots, N.$$

Thus, problem (3.34)–(3.37) is a difference formulation of the unsteady differential problem (3.22)–(3.25).

To increase stability, the dissipative terms are taken out to the upper time level and the condition of stability for scheme (3.34) is $\left(\max_{j,n} c\right) \tau/h < 1$, where $c = [(\gamma p + B_y^2 + B_z^2 + B_0^2 \sin^2 \theta)/V]^{1/2}$ is the magneto-sound velocity in the Lagrangian coordinates. Scheme (3.34) has been verified by calculating simple limiting cases, by mesh refinement, and by the fulfilment of the law of total energy conservation. In typical cases the value of the total energy has remained constant to the accuracy of 0.1% during $\approx 10^3$ steps in time.

Since the most thoroughly studied experiment are waves propagating across the magnetic field, let us first of all consider the numerical results for this very case ($\theta = 0$). At the Mach numbers $M < M_*$ and the initial plasma concentrations $n_0 \approx 10^{12}$ cm^{-3}, the basic role in forming a wave belongs to the dispersive effects, as the time of the current effect in the wave front is insufficient for effective development of the beam instability and, consequently, plasma turbulization due to the development of ion-sound instability is impossible. As a result, a shock wave with the oscillatory structure is formed [24].

When unsteady shock waves with the Mach numbers $M < M_*$ are excited in a more dense plasma with the initial concentration $n_0 \approx 10^{13}$ cm^{-3}, their structure gets more monotonic, since the conditions of exciting the beam and ion-sound instabilities are fulfilled. A sufficiently great collision frequency ν_{ef} results in a dissipative breaking of the oscillatory structure of the wave. A quasi-steady shock wave with the width of the forefront of the order of the dispersive length δ_e is formed. With falling initial temperature of the plasma, the criterion of the ion-sound instability excitation holds at less times from the beginning of the magnetic piston motion than was the case at greater initial temperatures; the domain of the developed ion-sound instability expands, the magnetic field profile becomes more monotonic, the forefront width increases. As shown by calculation [7], the collision frequency ν_{ef} within and behind the front of the wave is approximately two orders greater than the Coulomb collision frequency.

With growing initial plasma density n_0 the energy dissipation increases, and the dispersive effects are completely damped by the dissipative ones. In this case the profile of quasi-steady shock waves has the following characteristic domains:

(1) the 'piston' domain, associated with the magnetic field diffusion into the plasma as far as $\delta_1 \approx (t/\mu_0 \sigma)^{1/2}$. Here a continuous transition from the maximum magnetic field at the plasma–vacuum interface to a value equal to the shock wave amplitude takes place; the density increases from zero to its maximum value;

(2) the domain of the transition piston–forefront, wherein the magnetic field, density and temperature are constant;

(3) the shock wave front with the width $\Delta \approx \left(\mu_0 \sigma V_A (M - 1)\right)^{-1}$ equal to the dissipative length.

The density profile falls behind the magnetic field profile as fas as $\delta_2 \approx \Delta$, which is true for the case when the mechanism of dissipation in a shock wave front is conditioned by plasma resistivity. In the case of greater values of the external magnetic field amplitudes ($A \gtrsim 7$) no shock wave is formed and the plasma is 'raked' by the magnetic piston. Comparison of the solutions of the unsteady problem at moderate Mach numbers ($M \lesssim 2.5$), obtained with and without heat conductivity allowed for, has shown that account taken of heat conductivity results in an insignificant increase in the front width of the shock wave, i.e. at moderate Mach numbers the heat conductivity effect on the structure of a shock wave front can be neglected.

Therefore, in the wave front a turbulent domain arises, wherein the effective plasma conductivity $\sigma_{ef} = ne^2/m_e v_{ef}$ is considerably less than the classical conductivity $\sigma = ne^2/m_e v_e$; the anomalous resistivity results in collisionless energy dissipation in the wave front.

Now let us consider the results of solving the unsteady problem for the Mach numbers $M > 2.5$, beginning with a brief analysis of the solutions with no account taken of the electron heat conductivity. With growing amplitude of the magnetic field at the plasma–vacuum interface, both the velocity and the amplitude of a shock wave grow. For instance, in the case when the amplitude of the external magnetic field $A = 2.6$, the shock wave without heat conductivity allowed for is unsteady: its velocity and front width change with growing time. Moreover, unlike the case of quasi-steady conditions, even at slight changes in the front width of the magnetic field the steepness of the particle density profile continually grows — the particle density distribution approaches a discontinuous one. This is the case at the Mach numbers $M > M_* \approx 2.76$.

Heat conductivity accounted for changes in the pattern. At the value of the magnetic field behind the shock wave front $B_z \lesssim 3B_0$, the steepness of the density profile remains constant for some time after the piston was switched on, and the density profile can be characterized by an approximately constant width $\Delta_n \sim \chi$. At great amplitudes of the magnetic field ($B_z \gtrsim 3B_0$) the steepness of the density profile continuously increases, while the steepness of the magnetic field profile and the wave velocity remain nearly constant. The above facts make it possible to obtain in calculations at $B_z \lesssim 3B_0$ the quasi-steady conditions, under which the non-linear 'twisting' of the density profile is compensated for by the 'smearing' effect of heat conductivity, as well as the conditions of breaking the wave at $B_z > 3B_0$.

The front width of the magnetic field Δ_B greatly exceeds that of the density Δ_n, and we can suggest the existence of an isomagnetic jump of density (at a practically constant magnetic field). Similar results have been obtained at the values of the amplitudes of the external magnetic field $2.7 \leq A \leq 4$ and the shock wave amplitude $2.9B_0 \leq B_z \leq 3.0B_0$. At a further increase of the external field ($A > 4$), both the amplitude and velocity of the shock wave continuously grow due to unsteadiness associated with the magnetic piston, which results in the shock wave breaking (its approaching the stage of overturning).

The critical Mach numbers $3.4 < M_* < 3.5$, characterizing a shock wave in the presence of heat conductivity and magnetic viscosity at the moment of breaking (overturning), have been obtained for the values of the magnetic field amplitude at the plasma–vacuum interface $5 \leq A \leq 8$; the critical amplitude $B_z^* \approx 3B_0$. These values are in good agreement with the solutions of the steady problem [29].

Thus, the unsteady problem solution at $\theta = 0$ leads one to the following conclusions: (1) at the values of the Mach numbers $2.8 \lesssim M \lesssim 3.3$ and in the presence of electron heat conductivity there exists a quasi-steady isomagnetic jump in density; (2) at the Mach numbers $3.4 < M < 3.5$ there takes place breaking of shock waves.

If the angle θ between the plane of the shock wave front and the direction of the undisturbed magnetic field B_0 is other than zero, then an increase in θ conditions the following transformations in the shock wave structure. At $0 \leq \theta \ll (m_e/m_i)^{1/2}$ the dispersive effects related to the electron inertia predominate, and, hence, the wave front is followed by a number of damping oscillations the spatial size of which is equal to the length of dispersion $\delta_e = c/\omega_{0e}$ by the order of magnitude. Since at $\theta \lesssim (m_e/m_i)^{1/2}$ the dispersive effects associated with both the electron inertia and plasma anisotropy are pronounced, a shock wave has a complex structure, consisting of the oscillations behind the main front with the characteristic spatial scale δ_e, and those in front of the main front with the characteristis scale $c\theta/\omega_{0i}$. At last, at $\theta \gg (m_e/m_i)^{1/2}$ the dispersive effects associated with the plasma anisotropy are predominant, and the shock wave gets the characteristic features of oblique waves — an oscillatory precursor and the absence of oscillations behind the main front.

Now let us consider the oblique shock waves. As shown by calculations [7, 11] under the conditions when the anomalous resistivity does not develop, the wave has a pronounced oscillatory precursor the length of which increases with growing time due to the formation of new oscillations, so the wave is not steady. If there exists a dissipation induced by collisions of a collective nature, the profile of the magnetic field in the wave changes as compared to the purely Coulomb case: a quasi-steady shock wave is formed, the oscillation frequency in the precursor essentially decreases and the precursor length is approximately constant. The width of the transition domain Δ, including the oscillatory precursor, is in good agreement with the theoretical estimate $\Delta \approx M^{-1}\theta^2 \mathscr{H}$. Within the oblique wave front, there appears the y-component of the magnetic field, and also of an oscillatory structure. With growing density n_0 and according to dissipation, the shock wave profile changes from oscillatory to monotonic; the oscillations are nearly completely damped when the dissipative size $\delta_d \approx$ $\left(\mu_0 \sigma_{ef} V_A (M-1)\right)^{-1}$ becomes greater than $\delta \approx c\theta/\omega_{0i}$.

An increase in the amplitude of the magnetic field on the plasma–vacuum interface results in great unsteadiness of the process, an increase in the wave velocity and in essential changes in the shock wave structure as compared to the quasi-steady conditions considered above. These changes are as follows: (1) with growing time the earlier formed oscillations of the magnetic field disappear and the front becomes aperiodic; besides, there is a certain increase in the width of the main jump of the z-component of the field; (2) the steepness of the density profile (as well as of the longitudinal velocity of the particles) sharply increases with growing time, while the profile width of the magnetic field remains approximately constant — an isomagnetic jump.

These results indicate the approaching of an unsteady oblique shock wave at large Mach numbers to the phase of overturning. The critical Mach numbers $M_*(\theta)$, at which the oblique waves break down, obtained from the numerical calculations of unsteady shock waves, coincide within 2–3% with $M_*(\theta)$ for the stationary case [29]. As shown in [17, 45], until the very moment of turning over the basic experimental results on oblique waves are in a good agreement with our calculations.

The shock waves, propagating along the magnetic field B_0 at the frequencies $\omega \lesssim \Omega_B = eB_0/m_i$, and known as the switch-on waves, have been studied experimentally [67, 82]. Let us consider the numerical results for the system of equation (3.22) at $\theta = 90°$ [16, 32]. In line with the law of dispersion, the z-component of the magnetic field is of an oscillatory structure both in front of the main jump and behind it, since a switch-on shock wave is a superposition of two waves, the faster of which, called unusual, forms a precursor, while the slower one, called usual, forms an oscillatory train. The spatial scale of all the oscillations of the order of the dispersion length is $\delta_i = c/\omega_{0i}$, but in the falling behind train the oscillations are of less amplitude and size than in the precursor. The y-component of the magnetic field, arising in the wave, is also of an oscillatory structure, the phase shift between B_y and B_z in the precursor comprising 90°. The direction of rotation of the transversal magnetic field vector $B_\perp$ in front of the main front coincides with the direction of rotation of the electron in an undisturbed magnetic field B_0, while behind the main front it coincides with the ion direction in the field B_0. The width of the fore-front of the magnetic field profile $B_\perp = (B_y^2 + B_z^2)^{1/2}$ is much greater than that of the density profile. At a sufficiently low dissipation the number of oscillations and the density profile width increase with growing time, though the velocity and the width of the magnetic field profile remain nearly constant in the formed wave. An increasing level of dissipation results in the possibility of forming a quasi-steady switch-on wave, the width of the magnetic field profile and the density remaining practically unchanged.

As shown in [29], in strong steady switch-on shock waves with $M = M_* = 1.63$ there takes place an isomagnetic jump in density — a discontinuity of the gas-dynamical functions n, u at a continuous magnetic field. Solving the unsteady problem (3.22) at sufficiently great values of the piston magnetic field makes it possible to detect changes in the structure of a steady switch-on wave and the approach of this wave to the stage of overturning at $M \to M_*$. Comparing the numerical results with experiments [67] demonstrates that there exists a qualitative agreement of the data on the structure and velocity of the switch-on shock waves at $M < M_*$. A detailed comparison, however, is difficult to carry out, since in the above laboratory experiments the conditions were not purely one-dimensional; besides, the conditions considered by us with $M \to M_*$ have not been studied experimentally at all.

Thus, the analysis of the results of a bulky series of calculations of the unsteady problem (3.22) shows that:

(1) at a low level of dissipation and moderate Mach numbers, shock waves have a pronounced oscillatory character. In this case at $0 \le \theta \ll (m_e/m_i)^{1/2}$ behind the front there is a number of oscillations with a spatial scale of the order of δ_e (the dispersion is induced by electron inertia); at $\theta \lesssim (m_e/m_i)^{1/2}$ the profile consists of oscillations with the scale δ_e behind the main front and of the oscillations with the scale $c\theta/\omega_{0i}$ in front of the main front; at $(m_e/m_i)^{1/2} \ll \theta < 90°$

a shock wave has only on oscillatory precursor; at $\theta = 90°$ the whole profile of the wave is of an oscillatory character.

(2) at $M < M_*(\theta)$, when the conditions for developing the collective processes are fulfilled and the dissipation level grows, a regular oscillatory structure of shock waves is violated. A further increase in the dissipation results in a total damping of the oscillations and in the origination of a quasi-steady shock wave with its front width equal to the dissipative size by the order of magnitude;

(3) at the values of the Mach numbers $M < M_*(\theta)$ a quasi-steady isomagnetic jump in density takes place;

(4) a further increase in the amplitude of the magnetic field and the Mach number of the shock wave $(M \to M_*(\theta))$ results in great unsteadiness of the process, continuous growth of the density profile steepness, which indicates the approach of the unsteady wave to the phase of overturning.

Let us now consider the numerical results of the problem on expansion of a plasma cylinder over the plasma background of a lesser density, which is described by the system of equations (3.27) and conditions (3.29), (3.31), (3.32) [21]. These data have been obtained through the difference scheme which is a natural generalization of scheme (3.24) for a cylindrical case. The calculations have been carried out at the following parameters: $r_0 = 50$ cm, $n_0 = 10^{12}$ cm^{-3}, $T_e^0 = 100$ eV, $T_i^0 = 10$ eV, $B_0 = 3 \times 10^{-3}T$, in which case $V_A^0 = 6 \times 10^4$ m/sec is the velocity scale, $t^0 = 8 \times 10^{-6}$ sec is the time scale. At the moment $t = 0$ the magnetic field is homogeneous throughout, the ion temperature is constant, the density and electron temperature are given by formulae (3.31) with the coefficients $A_\varrho = 10$, $A_T = 5$, $\alpha_\varrho = \alpha_T = 10^2$, $r_T = 0.1\,r_0$. A dimensionless pressure drop in the plasmoid and the background equals 220. Higher pressure in the centre of the domain induces the plasma motion to the periphery, which changes the initial configuration of the magnetic field.

A space–time dependence of the magnetic field is given in Fig. 11; the solid line corresponds to the variant with the anomalous conductivity accounted for (with the effective frequency of collisions ν_{ef}), the dashed line — to the Coulomb conductivity. The anomalous con-

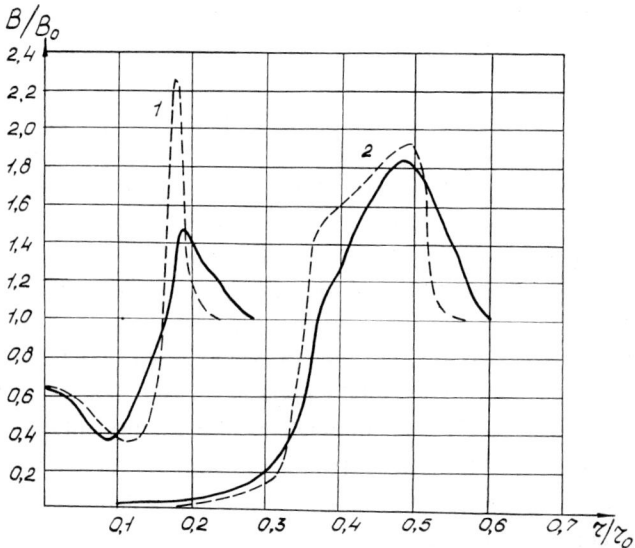

Fig. 11. Magnetic field profile in the process of a plasma cylinder expansion at successive moments of time. Dashed lines show the profiles with the Coulomb collisions allowed for, solid lines are those with the anomalous collisions taken into account. $1 - t = 0.02t_0$, $2 - t = 0.1t_0$.

ductivity, related to the beam and ion-sound instabilities, essentially reduces the magnetic field amplitude, especially at the initial moments of time, as the anomalous conductivity is greater than the Coulomb conductivity and, accordingly, the coefficient of the magnetic field diffusion is greater. With growing time the domain of the ion-sound instability development decreases, and the magnetic field amplitude at the anomalous conductivity differs but slightly from the case when only the Coulomb conductivity is allowed for. This fact can be accounted for by reducing the magnetic field gradients in the process of the plasma cylinder expansion. Note, that the plasma conductivity in the domain of the development of ion-sound instability is approximately 4 orders of magnitude less than the Coulomb conductivity.

To control the validity of the solution use was made of the law of the total energy conservation

$$\mathscr{W} = \int\limits_0^{r_0} \left[\frac{1}{2} nm_i u^2 + nm_e(\mu_0 en)^{-2} \left(\frac{\partial B}{\partial r}\right)^2 + \frac{3}{2} n(T_i + T_e) \right] r \, dr .$$

The $\mathscr{W}$ value changed in the calculations by 0.2% maximum.

The calculation analysis reveals that the heat energy of the electron gas first reduces and partially transfers into the energy increase of the magnetic field and the kinetic energy of the plasma; the heat energy of the ion gas also increases to some extent. Beginning from the time $t \simeq 0.16t^0$ the energy of every type remains approximately constant. At only the Coulomb conductivity allowed for, the time dependences of $\mathscr{W}_M, \mathscr{W}_{T_e}, \mathscr{W}_{T_i}, \mathscr{W}_{kin}$ remain unchanged, with the only exception of a slightly greater heating of the ion gas. Both at the Coulomb and anomalous conductivities, the magnetic field energy increases by some 20% as compared to the initial value. At the moments of time close to the beginning of expansion, the plasma cylinder velocity (i.e. the velocity of shifting of the forefront of the density profile) is approximately $7V_A^0$ at the anomalous conductivity and $5V_A^0$ at the Coulomb conductivity. Beginning from $t \simeq 0.1t^0$ the velocity of expansion remains constant in both cases and is $\approx 3V_A^0$.

Now let us consider the two-dimensional wave processes in a rarefied plasma without the supposition of smallness of the non-linear and dispersive effects by examining the problem of by-passing the conducting cylinder by a supersonic flow of the collisionless non-isothermal plasma [33]. As has been noted in Chapter 1, such a plasma is a dispersive medium for ion-sound waves, which can be described with the following equations:

$$\frac{\partial n_i}{\partial t} + \text{div} (n_i \boldsymbol{u}) = 0,$$

$$\frac{\partial \boldsymbol{u}}{\partial t} + (\boldsymbol{u}\nabla) \, \boldsymbol{u} = -(e/m_i) \, \nabla \varphi - (n_i m_i)^{-1} \, \nabla p_i + \nu(n_i m_i)^{-1} \, \Delta \boldsymbol{u}, \quad (3.38)$$

$$\varepsilon_0 \, \Delta \varphi = e(n_e - n_i), \quad n_e = n_0 \exp (e\varphi/T_e), \quad p_i = n_i T_i.$$

Here ν is a certain constant viscosity, $T_e \gg T_i$.

Let us solve the problem in the polar coordinates r, θ in the domain $R_0 \leq r \leq R_1$, $\theta_0 \leq \theta \leq \pi$, where R_0 is the cylinder radius, R_1 is the external boundary of the calculation domain. As the initial data throughout the domain let us set the parameters of the on-coming undisturbed flow, i.e.

$$n_i(r, 0) = n_0, \quad u(r, 0) = u_0, \quad \varphi(r, 0) = 0 \qquad (3.39)$$

where n_0, u_0 are the values of the density and velocity of the plasma flow in the infinity. Under the above supposition this flow is super-sonic, i.e. $u_0 > c_s = (T_e/m_i)^{1/2}$. On the surface of the cylinder (an equipotential surface) we shall set the conditions

$$\varphi(R_0, t) = \varphi_0. \qquad (3.40)$$

Let us place the external boundary $r = R_1$ sufficiently far from the body, in which case the body potential φ_0 is screened by the plasma and the boundary condition for the potential can be chosen as:

$$\varphi(R_1, t) = 0. \qquad (3.41)$$

The boundary conditions for the hydrodynamical functions are set as follows:

$$u(R_0, t) = 0, \quad u(R_1, t) = u_0, \quad n_i(R_1, t) = n_0. \qquad (3.42)$$

The values of the potential φ_0 on the body, as well as those of the electron and ion temperatures T_e, T_i will be considered as given con-stants, as in the corresponding experiments any potential can be fed into the body, while constancy of the temperatures in the problems on ion-sound waves is a common assumption. Therefore, equations (3.38) together with the initial (3.39) and boundary (3.40)–(3.42) con-ditions are a mathematical formulation of the problem under discus-sion. In formulating the picture of by-passing taking account of the finite temperature T_i is of principal importance, since the ion pressure $p_i = n_i T_i$ conditions the medium elasticity with respect to compression and results in ion repulsion from the body surface.

Now let us write equations (3.38) and conditions (3.39)–(3.42) in the dimensionless variables:

$$\frac{\partial n_i}{\partial t} + \text{div}\,(n_i \boldsymbol{u}) = 0,$$

$$\frac{\partial \boldsymbol{u}}{\partial t} + (\boldsymbol{u}\nabla)\,\boldsymbol{u} + \nabla\varphi + (\alpha/n_i)\,\nabla n_i = (\nu/n_i)\,\Delta\boldsymbol{u}; \qquad (3.43)$$

$$\beta\,\Delta\varphi = \exp\,(\varphi) - n_i. \qquad (3.44)$$

Here $\alpha = T_i/T_e$, $\beta = (D/R_0)^2$; the distances are normalized with respect to the cylinder radius R_0, the times to the quantity R_0/c_s, the potential — to the quantity T_e/e. In this problem the dispersive effects are induced by the charge separation within the scales of the order of the Debye electron radius D, so that in system (3.43) the coefficient β is a dispersion parameter.

Let us consider the algorithm of the numerical solution of problem (3.43), (3.44). In the polar coordinates the equations of motion and continuity are as follows:

$$
\begin{aligned}
&\frac{\partial u}{\partial t} + u\frac{\partial u}{\partial r} + \frac{v}{r}\frac{\partial u}{\partial\theta} - \frac{v^2}{r} = -\frac{\partial\varphi}{\partial r} - \frac{\alpha}{n}\frac{\partial n}{\partial r} \\
&\quad + \frac{\nu}{n}\left(\frac{\partial^2 u}{\partial r^2} + \frac{1}{r}\frac{\partial u}{\partial r} + \frac{1}{r^2}\frac{\partial^2 u}{\partial\theta^2} - \frac{2}{\tau^2}\frac{\partial v}{\partial\theta} - \frac{u}{r^2}\right), \\
&\frac{\partial v}{\partial t} + u\frac{\partial v}{\partial r} + \frac{v}{r}\frac{\partial v}{\partial\theta} + \frac{uv}{r} = -\frac{1}{r}\frac{\partial\varphi}{\partial\theta} - \frac{\alpha}{nr}\frac{\partial n}{\partial\theta} \qquad (3.45) \\
&\quad + \frac{\nu}{n}\left(\frac{\partial^2 v}{\partial r^2} + \frac{1}{r}\frac{\partial v}{\partial r} + \frac{1}{r^2}\frac{\partial^2 v}{\partial\theta^2} + \frac{2}{r^2}\frac{\partial u}{\partial\theta} - \frac{v}{r^2}\right), \\
&\frac{\partial n}{\partial t} + u\frac{\partial n}{\partial r} + \frac{v}{r}\frac{\partial n}{\partial\theta} + n\frac{\partial u}{\partial r} + \frac{n}{r}\frac{\partial v}{\partial\theta} + \frac{nu}{r} = 0.
\end{aligned}
$$

Here u is the r-component of the velocity, v is its θ-component, $n \equiv n_i$.

In the calculation domain $R_0 \le r \le R_1$, $\theta_0 \le \theta \le \pi$ let us introduce the difference mesh $r_i = R_0 + ih_r$, $\theta_k = \theta_0 + kh_\theta$, $t^n = n\tau$, $i = 1, \ldots, I$, $k = 1, \ldots, K$, $n = 0, \ldots, N$ and write the mesh functions as

$n(r_i, \theta_k, t^n) = n^n_{ik}$, $u(r_i, \theta_k, t^n) = u^n_{ik}$. System (3.45) has been solved by the method of fractional steps [110] in two stages. In the first step one calculates the function values in a fractional step in time:

first step

$$\tilde{u} = u - \tau \left\{ u \frac{\partial u}{\partial r} + \frac{v}{r} \frac{\partial u}{\partial \theta} - \frac{v^2}{r} + \frac{\partial \varphi}{\partial r} + \frac{\alpha}{n} \frac{\partial n}{\partial r} - \frac{v}{n} \left(\frac{\partial^2 \tilde{u}}{\partial r^2} \right. \right.$$

$$\left. \left. + \frac{1}{r} \frac{\partial \tilde{u}}{\partial r} - \frac{\tilde{u}}{r^2} + \frac{1}{r^2} \frac{\partial^2 u}{\partial \theta^2} - \frac{2}{r^2} \frac{\partial v}{\partial \theta} \right) \right\},$$

$$\tilde{v} = v - \tau \left\{ u \frac{\partial v}{\partial r} + \frac{v}{r} \frac{\partial v}{\partial \theta} + \frac{uv}{r} + \frac{1}{r} \frac{\partial \varphi}{\partial \theta} + \frac{\alpha}{nr} \frac{\partial n}{\partial \theta} \right.$$

$$\left. - \frac{v}{n} \left(\frac{\partial^2 \tilde{v}}{\partial r^2} + \frac{1}{r} \frac{\partial \tilde{v}}{\partial r} - \frac{\tilde{v}}{r^2} + \frac{1}{r^2} \frac{\partial^2 v}{\partial \theta^2} + \frac{2}{r^2} \frac{\partial u}{\partial \theta} \right) \right\},$$

$$\tilde{n} = n - \tau \left(u \frac{\partial n}{\partial r} + \frac{v}{r} \frac{\partial n}{\partial \theta} + n \frac{\partial u}{\partial r} + \frac{n}{r} \frac{\partial v}{\partial \theta} + \frac{nu}{r} \right),$$

$$\tilde{u}, \tilde{v}, \tilde{n} = u^{n+\frac{1}{2}}, v^{n+\frac{1}{2}}, n^{n+\frac{1}{2}}, u, v, n = u^n, n^n, v^n;$$

in the second step one calculates the function values in an integer step in time:

second step

$$u^{n+1} = \tilde{u} - \tau \left\{ \tilde{u} \frac{\partial \tilde{u}}{\partial r} + \frac{\tilde{v}}{r} \frac{\partial \tilde{u}}{\partial \theta} - \frac{\tilde{v}^2}{r} + \frac{\alpha}{\tilde{n}} \frac{\partial \tilde{n}}{\partial r} \right.$$

$$\left. - \frac{v}{\tilde{n}} \left(\frac{\partial^2 \tilde{u}}{\partial r^2} + \frac{1}{r} \frac{\partial \tilde{u}}{\partial r} - \frac{\tilde{u}}{r^2} + \frac{1}{r^2} \frac{\partial^2 u^{n+1}}{\partial \theta^2} - \frac{2}{r^2} \frac{\partial v^{n+1}}{\partial \theta} \right) \right\},$$

$$v^{n+1} = \tilde{v} - \tau \left\{ \tilde{u} \frac{\partial \tilde{v}}{\partial r} + \frac{\tilde{v}}{r} \frac{\partial \tilde{v}}{\partial \theta} + \frac{\tilde{u}\tilde{v}}{r} + \frac{\alpha}{\tilde{n}r} \frac{\partial \tilde{n}}{\partial \theta} \right.$$

$$\left. - \frac{v}{\tilde{n}} \left(\frac{\partial^2 \tilde{v}}{\partial r^2} + \frac{1}{r} \frac{\partial \tilde{v}}{\partial r} - \frac{\tilde{v}}{r^2} + \frac{1}{r^2} \frac{\partial^2 v^{n+1}}{\partial \theta^2} + \frac{2}{r^2} \frac{\partial \tilde{u}}{\partial \theta} \right) \right\},$$

$$n^{n+1} = \tilde{n} - \tau \left(\tilde{u} \frac{\partial \tilde{n}}{\partial r} + \frac{\tilde{v}}{r} \frac{\partial \tilde{n}}{\partial \theta} + \tilde{n} \frac{\partial \tilde{u}}{\partial r} + \frac{\tilde{n}}{r} \frac{\partial \tilde{v}}{\partial \theta} + \frac{\tilde{n}\tilde{u}}{r} \right).$$

All the terms of type $u \, \partial f/\partial r$ are approximated by the scheme of the first order of accuracy, allowing for the velocity direction. This scheme is monotonic and stable at $\tau \leq \min\limits_{i,k} [h(|u| + |v| + c_s\sqrt{2})^{-1}]$. The property of monotonicity is, generally speaking, extremely important in choosing a scheme for numerical calculations of dispersive media, since non-monotonic schemes induce their own 'numerical' dispersion, which is sometimes distinguished from the real physical dispersion only with difficulty.

In every time step the obtained equations have been solved by the method of scalar sweeping, the condition of the method stability fulfilled at any steps τ, h_r, h_θ.

Calculations have been carried out only in a half of the domain, as the problem is symmetrical with respect to the stagnation line. The problem symmetry was allowed for in the following way: two additional levels with respect to the angle θ were introduced and the function values at the additional nodes in the nth step in time were made use of, followed by introducing the values from the symmetrical nodes into the additional nodes.

For the flow to be completely determined, it is necessary to set the initial data of all the quantities throughout the domain, the boundary conditions of all the functions in the on-coming flow and the boundary conditions on the body surface. As the initial data we set the parameters of a homogeneous on-coming flow in the domain $i = 2, ..., I$, $k = 1, ..., K$: $u_{ik}^0 = u_0 \cos \theta_k$, $v_{ik} = -u_0 \sin \theta_k$. On the body surface the condition $u_{ik}^n = v_{ik}^n = 0$ $(i = 1, k = 1, ..., K)$ is set. To determine the density of the body surface use is made of the fact that both velocity components turn to zero on the body surface. In this case the calculation scheme is essentially simplified and makes it possible to calculate the density at the mesh nodes with the formula

$$n_{1,k}^{n+1} = n_{i,k}^n - \tau n_{1,k}^n u_{2,k}^n / h_r \quad (k = 1, ..., K).$$

Calculations have been carried out both in the complete $(\theta_0 = 0)$ and in the incomplete $(\theta_0 = 1)$ domain. In the latter case the boundary condition $(\partial f/\partial \theta)_{\theta = \theta_0} = 0$ is added.

Let us consider an algorithm of the numerical solution of the Poisson equation (3.44) with the boundary conditions $\varphi(R_0, t) = \varphi_0$, $\varphi(R_1, t) = 0$. To solve the equation the method of quasi-linearization, suggested in [5] is used, i.e.

$$\beta \Delta \varphi^{s+1} = (1 + \varphi^{s+1} - \varphi^s) \exp (\varphi^s) - n.$$

The Laplacian in the cylindrical coordinates $\Delta \varphi = \dfrac{1}{r} \dfrac{\partial}{\partial r} \left(r \dfrac{\partial \varphi}{\partial r} \right)$

$+ \dfrac{1}{r^2} \dfrac{\partial^2 \varphi}{\partial \theta^2}$ is replaced by the difference equation

$$\frac{1}{rh_r} \left(r_{i+\frac{1}{2}} \frac{\varphi_{i+1,k} - \varphi_{ik}}{h_r} - r_{i-\frac{1}{2}} \frac{\varphi_{ik} - \varphi_{i-1,k}}{h_r} \right)$$
$$+ (\varphi_{i,k+1} - 2\varphi_{ik} + \varphi_{i,k-1}) (r_i h_\theta)^{-2}.$$

If we put the points (i, k) in such an order that the point (i, k) precedes the point (i', k'), when $i < i'$, $k < k'$, and denote the obtained set $I \times K$ of the numbers $\{\varphi_{ik}^s\}$ through φ^s, then the method of quasi-linearization results in the necessity of solving a system of linear equations. Normalizing these equations by dividing all the terms into the coefficient at the central term φ_{ik}, we reduce the system to

$$(E - U_s - L_s) \varphi^{s+1} = f_s.$$

Here E, U, L are the unit, upper and lower matrices, respectively, of the size $(I \cdot K) \times (I \cdot K)$. The components of the vector f^s are equal to the corresponding values, assumed by the right-hand parts of equation (3.44).

To solve the equation the iteration method of the successive over-relaxation (SOR) has been chosen. The solution φ^{n+1} of equation (3.44) is found as a limit of the sequence of the solutions $\varphi^{n+1,s}$ of the linear equation

$$\beta \Delta \varphi^{n+1,s+1} = (1 + \varphi^{n+1,s+1} - \varphi^{n+1,s}) \exp (\varphi^{n+1,s}) - n^{n+1}.$$

In [5] they have proved a monotonic convergence of the sequence $\{\varphi^{n+1,s}\}$ to the solution of equation (3.44), in the case when $\varphi^{n+1,1} > \varphi^{n+1,0}$. Let us prove that $\varphi^{n+1,1}$ exceeds $\varphi^{n+1,0}$ at any initial approximation. The first approximation is found from the equation

$$\beta \Delta \varphi_1 = f(\varphi_0, r) + f_\varphi(\varphi_0, r) \cdot (\varphi_1 - \varphi_0), \qquad (3.46)$$

where $\varphi_0 = \varphi^{n+1,0}$, $\varphi_1 = \varphi^{n+1,1}$, f is the right-hand part of equation (3.44). Expanding the right-hand part of the equation by the Lagrange formula in the vicinity of the function $\varphi_0(r)$, we get:

$$\beta \Delta\varphi = f(\varphi_0, r) + f_\varphi(\varphi_0, r) \cdot (\varphi - \varphi_0) + \frac{1}{2} f_{\varphi\varphi}(\varphi_0, r) \cdot (\varphi - \varphi_0)^2.$$

$$(3.47)$$

Subtracting (3.46) from (3.47) and introducing the denotation $\psi = \varphi_1 - \varphi$, we have

$$\Delta\psi - f_\varphi(\varphi_0, r) \cdot \psi = -\frac{1}{2} f_{\varphi\varphi} \psi^2 \leq 0 \qquad (3.48)$$

since $f_{\varphi\varphi} > 0$. Let us demonstrate that for any point, belonging to the domain, $\psi \geq 0$. For this purpose let us assume that there exists such a point wherein $\psi < 0$. Since $\psi(R_0) = 0$, $\psi(R_1) = 0$, in the domain there is a point r_1, where $\psi(r_1)$ reaches its minimum. At this point $\Delta\psi(r_1) > 0$, $\psi(r_1) < 0$, which runs counter to inequality (3.48), thus proving our statement.

In every time step the solution of equation (3.44) in the preceding moment of time φ^n is used as an initial approximation of $\varphi^{n+1,0}$. In the finite differences we have

$$\varphi_{ik}^{s+1} = (A_i + B_i + C_i + \exp(\varphi_{ik}^s)) - \omega(B_i \varphi_{i-1,k}^{s+1}$$
$$+ C_i \varphi_{i,k-1}^{s+1}) = \omega(A_i \varphi_{i+1,k}^s + C_i \varphi_{i,k+1}^s)$$
$$- \varphi_{ik}^s(\omega - 1)(A_i + B_i + C_i + \exp(\varphi_{ik}^s))$$
$$+ \omega[n_{ik}^{n+1} - \exp(\varphi_{ik}^s) \cdot (1 - \varphi_{ik}^s)],$$

$$(3.49)$$

where ω is the relaxation parameter.

The criterion for the iteration process cessation is based on the fact that the maximum of the absolute value of the relative change in the function value for two subsequent iterations must not exceed a certain small number ε, i.e. $\max_{i,k} |\varphi_{ik}^{s+1} - \varphi_{ik}^s| < \varepsilon$, in which case $\varphi_{ik}^{s+1} = \varphi_{ik}^{n+1}$.

The trial calculations carried out by the difference scheme (3.49) made it possible to choose the optimal parameter of relaxation $\omega = 1.7$, the viscosity coefficient $\nu = 0.01$, which ensures the dispersive oscilla-

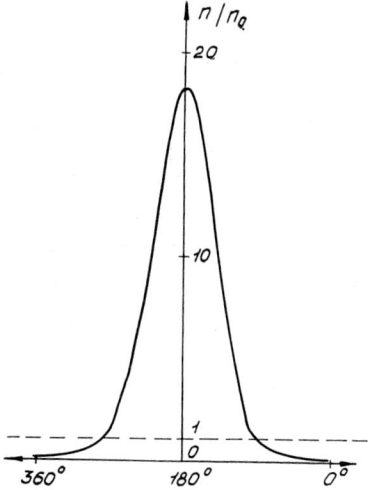

Fig. 12. Angular dependence of ion density on a cylinder surface.

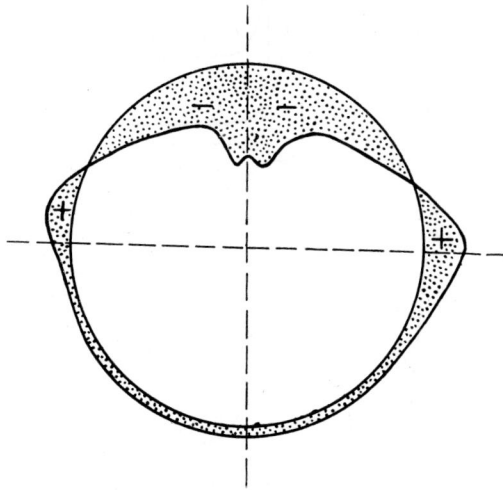

Fig. 13. Angular dependence of ion density disturbances at the distance $r = 2R_0$ from the cylinder surface.

tions damping and the shock wave formation, and the external radius of the calculation domain $R_1 = 3R_0$.

To clarify a general picture of the flow, calculations have been carried out in the complete domain $R_0 \leq r \leq 3R_0$, $0 \leq \theta \leq \pi$ for the following plasma parameters: $u_0 = 1, 2$, $\beta = 0.09$, $\alpha = 0.1$, $\varphi_0 = 0.5$ (in the dimensionless variables). Figure 12 presents the angular dependence of the ion density $n(r, t)$ on the body surface at the moment of time $t = 1.5$. A domain of compression $n_{\max} \approx 18.5$ is formed in front of the body ($\theta = 180°$), and a rarefied prolonged wake ($n_{\min} \simeq 10^{-4}$) behind the body. Figure 13 demonstrates the angular dependence of the ion density disturbances $\delta n = n - 1$ as far as $r = 2R_0$ from the cylinder surface at the moment of time $t = 1$, the circumference $r = 2$ corresponds to the undisturbed density $n = 1$;

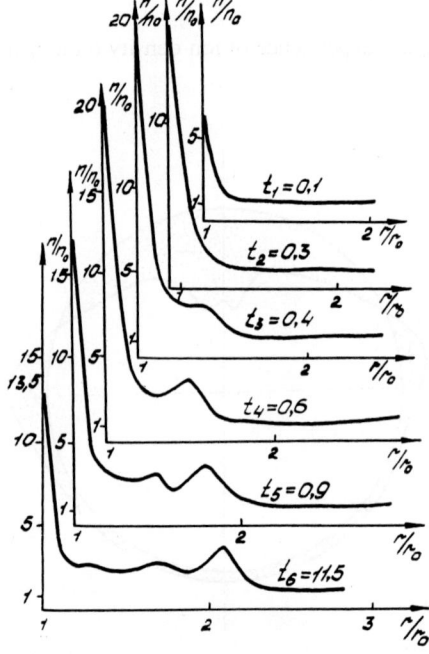

Fig. 14. Ion density profile on the stagnation line at successive moments of time.

at $r > 2$ the values $\delta n > 0$ (compression) are plotted, at $r < 2$ — the values $\delta n < 0$ (rarefaction). Behind the cylinder one can see a rarefied wake with a characteristic ion focusing around the axis, coinciding with the direction of the on-coming plasma flow. An analogous effect of ion focusing also takes place in calculating flows around bodies in the kinetic approach [3].

In the vicinity of the body a shock wave with a dispersive structure arises. As is seen in the profile of the ion density on the stagnation line (Fig. 14), in successive moments of time the ion density at the body surface (moments of time $t_1 = 0.1$, $t_2 = 0.3$) first grows to the values $n \approx 20$. Then, as a sufficient density is accumulated at the body surface, the ion pressure comes into play, pushing the ions away from the surface (the moment $t = 0.4$), and a wave of compression gets started from the body. In the course of the initial disturbance propagation in a dispersive medium, the disturbance gradually assumes a characteristic dispersive shape (the moment of time $t = 0.6$), and begins to go away from the body (the moments of time $t = 0.9$, $t = 1.15$).

Chapter 4

Shock Waves in a Rarefied Plasma (hybrid models)

As has been shown in the preceding chapter, with the growing amplitude of a shock wave the steepness of the density profile and the potential jump sharply increases. Plasma instabilities induce anomalous resistivity and heat conductivity, associated with collective interactions, but do not induce viscosity, since the experiments [46, 47, 49, 68] have always demonstrated lagging of the density profiles from those of the magnetic fields, a fact which indicates that the character of the shock wave front is of a resistive (conditioned by the finite conductivity) rather than of a viscous nature, as in this case the density and the magnetic field profiles would be close to one another. That is why the phenomenon of turning over of the large-amplitude waves is possible in a collisionless plasma [93]. Studying strong shock waves immediately before, in the process of and after the turning over cannot be carried out based on the models of a gas-dynamical type, as they become incorrect due to a multi-current flow domain arising at turning over.

In a plasma with the magnetic field it is necessary to calculate the motion of both ions and electrons. But a completely kinetic modelling of ions and electrons is difficult to realize, as there is too great a difference in spatial and time scales of the ion and electron motion. To study such cases, there have been developed and are widely used combined (hybrid) numerical models, wherein the ion plasma component is described within the kinetic approach, while the electron component — by the equations of a gas-dynamical type [31, 41, 57, 100]. Due to the smallness of the electron Larmor radius $r_L = v_{Te}/\omega_B$ (v_{Te} is the electron thermal velocity), the limits of the gas-dynamical approach are much wider for electrons than for ions, since the motion of the latter in the process under discussion obeys no gas-dynamical (one-velocity) description. The gas-dynamical approach for electrons makes it possible to eliminate the unessential for these phenomena small spatial scales $\sim D$ and high frequencies $\omega \sim \omega_{0e}$, related to electrons, and adjust the mathematical model for computer realization.

In the problems on shock waves the gas-dynamical description for ions is violated only in the domains with large gradients of the plasma and field characteristics, and, hence, calculations of the ion component motion can be simplified in the following way. In 'undisturbed' regions, where the gas-dynamical description can be used, we make use of the equation of plasma as a continuous medium (of type (3.2) and the corresponding numerical algorithms, while in the domains with strong gradients of the plasma and field parameters — of the method of particles for solving the Vlasov kinetic equation for ions, and the gas-dynamical equations for electrons.

Let us begin considering hybrid models with the description of the algorithmic aspects of the method of particles. A plasma is viewed as a set J of the particles located in the calculation volume V. The position of every particle with the charge e_j and the mass m_j in the phase space is characterized by the radius-vector $r_j(t)$ and velocity $v_j(t)$. The distribution function of the point particles

$$f(r, v, t) = \sum_{j=1}^{J} \delta(r - r_j(t))\, \delta(v - v_j(t))$$

can be used in calculations, but with difficulty, as there can arise significant density fluctuations due to the limited number of particles and, besides, there takes place a divergence of the electric field at small distances. In order to reduce the level of fluctuations a modification of the method of particles has been suggested under the assumption that the particles have a spatial size and can freely pass through one another [35]. If the particles have a finite width, at small distances the forces are smoothed down, while the collective interactions, playing the basic part, are described sufficiently well. The distribution function for the particles of a finite size can be written [78, 99] as

$$\tilde{f}(r, v, t) = \int f(r', v, t) \, R(r', r) \, dr'$$

$$= \sum_{j=1}^{J} R(r, r_j(t)) \, \delta(v - v_j(t)), \qquad \int_V R(r) \, dr = 1,$$

(4.1)

from which one can find the expressions for the density of the number of particles and the current density

$$\tilde{\varrho}(r, t) = \sum_{j=1}^{J} e_j R(r, r_j(t)),$$

$$\tilde{j}(r, t) = \sum_{j=1}^{J} e_j v_j(t) \, R(r, r_j(t)).$$

The kernel R in transform (4.1) characterizes the geometrical shape, particle size and density distribution inside the particle. The kernel R can, in principle, be of any kind provided it fulfils its basic task — to smooth down the forces and to eliminate the close interactions which are of minor importance in a rarefied plasma. The necessity of a simple realization and approximation, however, results in using in practical calculation a limited number of simplest kernels — a rectangle, triangle, Gaussian. For instance,

$$R(x, x') = \begin{cases} (2\Delta)^{-1}, & |x - x'| \leq \Delta \\ 0, & |x - x'| > \Delta \end{cases}$$

describes a particle with the width 2Δ and a uniform charge distribution, while the kernel

$$R(x, x') = \begin{cases} \Delta^{-1}(1 - |x - x'|/\Delta), & |x - x'| \leq \Delta \\ 0, & |x - x'| > \Delta \end{cases}$$

describes a particle with the width 2Δ and a non-uniform distribution of the charge density.

If the coordinates and particle velocities are given, then, making use of the kernels $R(r, r')$, one can find the density of the charge $\tilde{\varrho}(r)$, current $j(r)$ throughout the domain, and on their basis, solving the

Maxwell equations, determine the values of the electric and magnetic fields also throughout the domain. For the majority of practically interesting problems an analytical solution of the Maxwell equations is impossible. Hence, use should be made of approximated finite-difference methods, in which case a spatial mesh is to be introduced. Here an important problem arises: in which way to approximate the charge and current density in the mesh nodes. The simplest way is to ascribe to every node the values of the functions $\tilde{\varrho}, \tilde{j}$ at this point, i.e. $\varrho_k = \tilde{\varrho}(r_k), j_k = \tilde{j}(r_k)$, where k is the node number. But in this case the error can be enormous, and, therefore, the charge density transfer to the spatial mesh nodes requires caution.

Let us now consider the calculation domain S where the particles, modelling a distribution of the density charge $\varrho(r)$, are located. Let us introduce in this domain a spatial mesh with the nodes r_k, wherein we shall determine ϱ_k. Let the domain S be covered with disjoint cells s_k in such a way that there is one node r_k inside every cell. Let us require the fulfilment of the natural equality, expressing the law of the total charge conservation:

$$\sum_k s_k \varrho_k = \int_S \varrho(r) \, dr$$

where ϱ_k is the charge density, which is to be ascribed to every node of the mesh r_k. The condition of the charge conservation is met if the ϱ_k values in the mesh nodes are determined through the procedure

$$\varrho_k = s_k^{-1} \int_{s_k} \tilde{\varrho}(r) \, dr. \tag{4.2}$$

Let us further limit ourselves by a rectangle domain S, wherein a Cartesian mesh with the steps h_1, h_2 along the directions x, y, respectively, has been introduced. In this case the cells s_k are rectangles with the sides h_1, h_2, in the centres of which the mesh nodes with the coordinates (x_i, y_k) are located. Formula (4.2), then, assumes the form

$$\varrho_{ik} = (h_1 h_2)^{-1} \int_{y_k - \frac{h_2}{2}}^{y_k + \frac{h_2}{2}} \int_{x_i - \frac{h_1}{2}}^{x_i + \frac{h_1}{2}} \tilde{\varrho}(x, y) \, dx \, dy. \tag{4.3}$$

Let us write the restored charge density in the Cartesian coordinates under the assumption that the transform kernel R is symmetrical:

$$\tilde{\varrho}(x, y) = \sum_j e_j R(x - x_j, y - y_j). \tag{4.4}$$

Substituting (4.4) into (4.3), we determine the charge density in the mesh nodes (mesh density):

$$\varrho_{ik} = \sum_j e_j \bar{R}(x_i - x_j, y_k - y_j) \tag{4.5}$$

where $\bar{R}$ is a mesh kernel related to the initial kernels R through the formula

$$\bar{R}(x, y) = (h_1 h_2)^{-1} \int_{-\frac{h_2}{2}}^{\frac{h_2}{2}} \int_{-\frac{h_1}{2}}^{\frac{h_1}{2}} R(x - x', y - y') \, dx' \, dy'.$$

Formulae (4.4) and (4.5) demonstrate that both the restored and mesh densities of the charge are calculated in the same way with the only exception: the initial kernel R is replaced by the mesh kernel $\bar{R}$.

The particle models used in practice can be classified by the type of the mesh kernel $\bar{R}$. Chronologically, the first and the simplest model is the NGP (Nearest-Grid-Point) model [56]. The charge of every particle located in the cell s_k that surrounds the mesh node r_k is ascribed to this node. The charge density ϱ_{ik} in the node (for the sake of definiteness the two-dimensional case is considered) is obtained by dividing the total charge of the particles trapped in the cell s_{ik} into the area of this cell. Such density restoration is realized through the kernels R, $\bar{R}$ described in [28].

Somewhat later, there appeared the modifications referred to in the literature as the PIC (Particle-in-Cell) model [78] and the CIC (Cloud-in-Cell) model [34]. Within the PIC approach, the charge of every particle is distributed through the inverse linear interpolation between the two nearest mesh nodes in the one-dimensional case and through the inverse bilinear interpolation among the four nearest mesh nodes in the two-dimensional case. Let a particle with the coordinate x_j and the charge e_j be located between the nodes x_{i-1}, x_i of the uniform mesh with the step h (the one-dimensional case). In this case this particle brings the charge $e_j h^{-1}(x_i - x_j)$ into the node x_{i-1}, and the charge $e_j h^{-1}(x_j - x_{i-1})$ into the node x_i; the charge density is obtained by dividing the sum of the charge contributions of the particles into the node on the mesh step h.

While considering the two-dimensional case let us assume that a particle with the coordinates x_j, y_j and the charge e_j is located among the nodes (x_{i-1}, y_{k-1}), (x_{i-1}, y_k), (x_i, y_k), (x_i, y_{k-1}) of a rectangle mesh with the steps h_1, h_2. The contribution of the particle charge into these nodes is given by the formulae:

$$e_j(h_1 h_2)^{-1} (x_i - x_j) (y_k - y_j) \quad \text{the node } (x_{i-1}, y_{k-1})$$

$$e_j(h_1 h_2)^{-1} (x_i - x_j) (y_j - y_{k-1}) \quad \text{the node } (x_{i-1}, y_k)$$

$$e_j(h_1 h_2)^{-1} (x_j - x_{i-1}) (y_j - y_{k-1}) \quad \text{the node } (x_i, y_k)$$

$$e_j(h_1 h_2)^{-1} (x_j - x_{i-1}) (y_k - y_j) \quad \text{the node } (x_i, y_{k-1}).$$

By way of dividing the sum of the particle contributions into every node into the cell area, we get the charge density in the node.

Within the CIC approach, the particles are viewed as rectangles with a uniform charge density; the sides $2\Delta_1$, $2\Delta_2$ are not related, in a general case, through any correlation with the spatial mesh steps h_1, h_2. The coordinates of the rectangle centre are considered the particle coordinates. The contribution of the particle charge into the mesh nodes is proportional to the part of the area belonging to the cell surrounding the corresponding node. If the mesh steps h_1, h_2 are fixed, the number of the nodes wherein the particle brings its charge depends, in a general case, on the particle sizes $2\Delta_1$, $2\Delta_2$ and its location [28].

The PIC and CIC models can be described in one way, using the same mesh kernel $\bar{R}$ given by formulae (2.23), (2.24) [28]. The difference between these modifications lies only in the fact that in the PIC model $2\Delta_1 = h_1$, $2\Delta_2 = h_2$, while in the CIC method the correlation between the sizes of the particle and the cell is arbitrary. Thus, the

PIC model is but a special case of the CIC model, the same being true for the NGP model, which is a CIC model modification at $\Delta_1, \Delta_2 \to 0$.

From the governing kinetic equation describing the particle dynamics, through an ordinary procedure one can easily get a system of equations for the moments of the distribution function:

$$\frac{\partial}{\partial t} \int v^n f \, dv + \frac{\partial}{\partial r_l} \int v^n v_l f \, dv - m^{-1} \int \left(\frac{\partial}{\partial v_l} F_l v^n \right) f \, dv = 0, \qquad (4.6)$$

where $n = 0, 1, \ldots$, and the summation is carried out with respect to the index l. Let us substitute the distribution function (4.1) for the particles of a finite size into equation (4.6) and obtain the following system of equations for the coordinates $r_j(t)$ and velocities $v_j(t)$ of the particles (see, for instance, [99]):

$$\sum_{j=1}^{J} \left[\dot{r}_j \frac{\partial R(r, r_j)}{\partial r_j} + v_j \frac{\partial R(r, r_j)}{\partial r} \right] v^n$$

$$+ m^{-1} \sum_{j=1}^{J} (m \dot{v}_{jl} - F_l) R(r, r_j) \frac{\partial v_j^n}{\partial v_{jl}} = 0. \qquad (4.7)$$

Let us integrate (4.7) with respect to the volume occupied by the system of particles under investigation. The resulting system of equations will be met in the case of the symmetrical kernels $R(r, r_j) = R(r - r_j)$ provided the following conditions are fulfilled:

$$\dot{r}_j(t) = v_j(t), \qquad (4.8)$$

$$m \dot{v}_j(t) = \int F(r, t) R(r - r_j) \, dr \equiv F_j(t), \quad j = 1, \ldots, J. \qquad (4.9)$$

Conditions (4.8), (4.9) are the equations of the particle motion. If these equations are solved accurately, then the moments of the distribution function for the particles of a finite size, integrated with respect to the volume, will be reproduced correctly, which is equivalent to a correct change in the physical quantities in the volume (mass, momentum, kinetic energy).

The integral in equation (4.9) is a force affecting the particle numbered j and described by the kernel R. Interpolation has to be carried out in this or that way, since the electric and magnetic fields in the expression for the Lorentz force are determined on a discrete set of points (nodes or cell centres of the spatial mesh), while for calculating $F_j(t)$ it is necessary to know the field distribution throughout the domain. Depending on the type of approximation, one gets various formulae for approximating the force at the particle location.

Let us express the force $F(r, t)$ through its values $F(r_k, t) = F_k(t)$ in the mesh nodes in the following way:

$$F(r, t) = \sum_k F_k(t) S(r - r_k)$$

where S is a weight function obeying the condition of normalization $\sum_k S(r - r_k) = 1$ for all the r values. The function S type governs the character of the force interpolation. Let us now consider a number of examples, for simplicity limiting ourselves to the one-dimensional case:

(1)
$$S(x - x_k) = \begin{cases} 1 \text{ at } |x - x_k| \leq h/2, \\ 0 \text{ at } |x - x_k| > h/2; \end{cases}$$

$$R(x - x_j) = \delta(x - x_j).$$

In this case

$$F_j = \int \sum_{k=1}^{K} F_k S(x - x_k) \, \delta(x - x_j) \, dx$$

$$= \sum_{k=1}^{K} F_k \int S(x - x_k) \, \delta(x - x_j) \, dx \qquad (4.10)$$

$$= \sum_{k=1}^{K} F_k S(x_j - x_k) = F_{k'},$$

where k' is the number of the mesh node, nearest to the particle with the coordinate x_j, i.e. $x_k' \in \left[x_j - \dfrac{h}{2}, x_j + \dfrac{h}{2} \right]$. Calculating the force by formula (4.10) coincides with the force definition by the method

1/0 suggested in [72].

(2)

$$S(x - x_k) = \begin{cases} 1 & \text{at } |x - x_k| \le h/2, \\ 0 & \text{at } |x - x_k| > h/2; \end{cases}$$

$$R(x - x_j) = \begin{cases} h^{-1} & \text{at } |x - x_j| \le h/2, \\ 0 & \text{at } |x - x_j| > h/2. \end{cases}$$

In this case

$$F_j = h^{-1} \sum_k F_k \int_{x_j-h/2}^{x_j+h/2} S(x - x_k) \, dx.$$

If $x_{k-1} < x_j < x_k$, then $F_j = h^{-1}[(x_k - x_j) F_{k-1} + (x_j - x_{k-1}) F_k]$. As is seen, the force affecting a particle is determined by linear interpolation between the field values in the two nearest to the particle nodes of the spatial mesh and coincides with the corresponding formulae of the calculations by the methods 1/1 and 2/1 suggested in [72].

(3)

$$S(x - x_k) = \begin{cases} 1 & \text{at } |x - x_k| \le h/2, \\ 0 & \text{at } |x - x_k| > h/2; \end{cases}$$

$$R(x) = \begin{cases} h^{-1}(1 - |x|/h) & \text{at } |x| \le h, \\ 0 & \text{at } |x| > h. \end{cases}$$

In this case

$$F_j = \sum_k F_k \int S(x - x_k) R(x - x_j) \, dx$$

$$= \sum_k F_k \int_{x_k - \frac{h}{2}}^{x_k + \frac{h}{2}} R(x - x_j) \, dx.$$

At $x_k - h/2 \le x_j \le x_k + h/2$ we get the following expression for the force:

$$F_j = (1/2h^2) \{ (x_j - x_k - h/2)^2 F_{k-1}$$
$$+ [(3/2) h^2 - 2(x_j - x_k)^2] F_k + (x_j - x_k + h/2)^2 F_{k+1} \}. \tag{4.11}$$

Let $x_j = x_k - h/2$ (the particle is located exactly between the nodes k, $k-1$), then the force affecting the particle is $F_j = \dfrac{1}{2}(F_{k-1} + F_k)$. If $x_j = x_k$ (the particle is located in the node), it is affected by the force $F_j = (F_{k-1} + 6F_k + F_{k+1})/8$, which, due to a certain 'smearing' of the particle charge over the mesh nodes, is not equal to the force F_k. Formula (4.11) shows that the force affecting the particle is determined by a quadratic interpolation.

As is seen from the formulae given above, in a quantitative sense calculations are essentially dependent on the type of the kernels R and S and are directly proportional to the number of particles. The more complex is the kernel, the greater quantity of calculations it requires. As a result, in practice only the simplest schemes of the force approximation are used.

A scheme of the multiple expansion [39, 66] for determining the forces affecting the particles requires, as compared to the described schemes, fewer operations per particle due to a significant increase in calculations per mesh node. Besides, in the scheme of the multipole expansion complication of the kernel used increases the number of operations per mesh node, while the number of operations per particle remains unchanged. Since in the method of particles the number of the particles greatly exceeds that of the nodes, the scheme in question can prove advantageous.

Let us consider the scheme of a multipole expansion through an example of a one-dimensional electrostatic problem with the electrons modelled by the particles, and the ions forming a motionless neutralizing background. In this case the density of the electron component is determined by the formula

$$\varrho(x) = e \sum_j R(x - x_j) \tag{4.12}$$

where R is an arbitrary kernel. Let the domain be subdivided into the similar cells of the size h, with the mesh nodes x_l located in the cell centres, and the function $R(x)$ be infinitely deferentiable. Then, by expanding the function $R(x)$ into a Taylor series, formula (4.12) can be transformed in the following way:

$$
\begin{aligned}
\varrho(x, t) = e \sum_l \sum_{j\in l} R(x_j(t) - x) &= e \sum_l \sum_{j\in l} [R(x_l - x) \\
&+ \Delta x_j R'(x_l - x) + (\Delta x_j)^2/(2!) R''(x_l - x) + \ldots] \\
- e \sum_l \Big[R(x_l - x)\, \varrho_0(l, t) &+ R'(x_l - x)\, \varrho_1(l, t) \\
&+ \frac{1}{2!} R''(x_l - x)\, \varrho_2(l, t) + \ldots \Big],
\end{aligned}
\tag{4.13}
$$

where

$$
\varrho_m(l, t) = \sum_{j\in l} (x_j(t) - x_l)^m
\tag{4.14}
$$

s the multipolar moment numbered m and calculated with respect to all the particles located in cell numbered l; $\Delta x_j = x_j(t) - x_l$ is the distance between the jth particle and the nearest mesh node x_l. Using a discrete Fourier transform, we get from (4.13)

$$
\varrho(k, t) = eR(k) \sum_{m=0}^{\infty} \frac{(-ik)^m}{m!} \sum_l \varrho_m(l, t) \exp(-ikx_l)
\tag{4.15}
$$

where $|kh| < \pi$. Thus, having determined the multipolar moments by formulae (4.14), we can find the amplitudes of the density harmonics by formulae (4.15). Then, employing the Poisson equation $\varepsilon_0\, \partial E/\partial x = -\varrho(x, t)$, we can easily determine the amplitudes of the electric field harmonics:

$$
E(k, t) = (i/\varepsilon_0 k)\, \varrho(k, t).
\tag{4.16}
$$

In line with formula (4.9), the force affecting the particle located at the point $x_j(t)$ is $F(x_j, t) = e \int E(x, t) R(x - x_j)\, dx = e \int E(x + x_j, t) R(x)\, dx$. Expanding the electric field under the sign of integral into a Taylor series and denoting through

$$
F_m(x_l, t) = e \int \frac{\partial^m E(x + x_l, t)}{\partial x^m} R(x)\, dx
$$

the mth moment of the force, we get

$$F(x_j, t) = F_0(x_l, t) + \Delta x_j F_1(x_l, t) + \frac{(\Delta x_j)^2}{2!} F_2(x_l, t) + \ldots \quad (4.17)$$

The amplitudes of the harmonics of the force moments $F^m(k, t)$ are related to those of the electric field harmonics through the formulae

$$F_0(k, t) = -e R(k) E(k, t), \quad F_m(k, t) = (ik)^m F_0(k, t). \quad (4.18)$$

Thus, having determined the amplitudes of the density harmonics $\varrho(k, t)$, one can find the amplitudes of the harmonics of the force moments $F_m(k, t)$ by formulae (4.16), (4.18) and then calculate the force affecting the particle numbered j by formula (4.17). It is common practice in schemes of multipolar expansion to choose the kernel of the type $R(x) = (2\pi)^{-1} \Delta^{-1/2} \exp(-x^2/2\Delta^2)$, the use of which in other schemes results in a highly ineffective algorithm of the particle motion, since one has to calculate the exponent many times.

The methods of numerical calculation of the equations of the particle motion in both fixed and obtained as a function of the coordinates and time electro-magnetic fields are given in [28].

Let us investigate the combined (hybrid) method for numerical studying large-amplitude unsteady shock waves propagating across the magnetic field in a rarefied quasi-neutral plasma. Let all the functions depend only on the coordinate x, the magnetic field be directed along the z-axis, and the direction of the wave propagation coincide with the x-axis. Let us write the equations of the electron gas motion and the equation of induction:

$$m_e \left(\frac{\partial u_e}{\partial t} + u_e \frac{\partial u_e}{\partial x} \right) = -e(E_x + v_e B) - \frac{1}{n} \frac{\partial}{\partial x}(nT_e) \quad (4.19)$$

$$m_e \left(\frac{\partial v_e}{\partial t} + u_e \frac{\partial v_e}{\partial x} \right) = -e(E_y - u_e B) - m_e \nu_{ef}(v_e - \langle v_i \rangle) \quad (4.20)$$

$$\frac{\partial B}{\partial x} = \mu_0 e n(v_e - \langle v_i \rangle).$$

Here u_e, v_e are the xth and yth components of the macroscopic velocity of the electron gas; $\langle v_i \rangle$ is the yth component of the macroscopic ion velocity; the term that is proportional to v_{ef} describes friction between the ions and electrons conditioned by collective interactions.

The equations of ion motion, which are the characteristics of the Vlasov kinetic equation, are as follows:

$$m_i \frac{du_i}{dt} = e(E_x + v_i B)$$

$$m_i \frac{dv_i}{dt} = e(E_y - u_i B) + m_e v_{ef}(v_e - \langle v_i \rangle) \qquad (4.21)$$

$$\frac{dx_i}{dt} = u_i.$$

Here x_i is the ion coordinate, u_i, v_i are the xth and yth components of the velocity of single ions. The term proportional to v_{ef} describes the momentum related to the collective interactions and transferred from ions to electrons (at one ion per time unit). This change in the momentum must be introduced in order to bring the discussed model in accord with the gas-dynamical model which is valid at small wave amplitudes.

Under the supposition, the plasma is quasi-neutral, so the x-components of the macroscopic velocity of electrons and ions are equal: $u_e = \langle u_i \rangle$. Equation (4.20) affords that $v_e = \langle v_i \rangle + (\mu_0 en)^{-1} \dfrac{\partial B}{\partial x}$.

Therefore, the macroscopic velocity of electrons are expressed through those of ions and through the magnetic field, and the equations of electron motion (4.19) allow one to determine the components of the electric field:

$$E_x = -v_e B - \frac{1}{en} \frac{\partial}{\partial x}(nT_e) - \frac{m_e}{e} \frac{d\langle u_i \rangle}{dt}$$

$$= -\langle v_i \rangle B - \frac{m_e}{e} \frac{d\langle u_i \rangle}{dt} - \frac{1}{en} \frac{\partial}{\partial x}\left(nT_e + \frac{B^2}{2\mu_0}\right) \qquad (4.22)$$

$$E_y = \langle u_i \rangle\, B - \frac{m_e \nu_{ef}}{\mu_0 e^2 n} \frac{\partial B}{\partial x} - \frac{m_e}{e} \frac{dv_e}{dt} \tag{4.22}$$

$$= \langle u_i \rangle\, B - \frac{m_e}{\mu_0 e^2} \left(\frac{\nu_{ef}}{n} \frac{\partial B}{\partial x} + \frac{d}{dt} \frac{1}{n} \frac{\partial B}{\partial x} \right) - \frac{m_e}{e} \frac{d\langle v_i \rangle}{dt}.$$

The equations for the magnetic field and for the electron temperature are as follows:

$$\frac{\partial B}{\partial t} = -\frac{\partial E_y}{\partial x}$$

$$\frac{3}{2} \left(\frac{\partial T_e}{\partial t} + \langle u_i \rangle \frac{\partial T_e}{\partial x} \right) = -T_e \frac{\partial}{\partial x} \langle u_i \rangle$$

$$+ m_e \nu_{ef} \left(\frac{1}{\mu_0 e n} \frac{\partial B}{\partial x} \right)^2 + \frac{1}{n} \frac{\partial}{\partial x} \left(\chi \frac{\partial T_e}{\partial x} \right). \tag{4.23}$$

Let at the initial moment of time the following state of plasma be given:

$$n(x, 0) = n_0 \left[1 + \frac{A - 1}{1 + \exp (x - x_0)\, l} \right] \tag{4.24}$$

$$B(x, 0) = (B_0/n_0)\, n(x, 0), \quad u(x, 0) = 0, \quad T_e(x, 0) = T^0 = \text{const}.$$

The quantity A determines the initial drop in density and magnetic field, the parameter l characterizes the initial profile steepness.

The domain $0 \le x \le L$, wherein a solution of the considered problem is being sought, is divided into three domains depending on the initial profile n, B and the expected character of the flow. As follows from the qualitative considerations, with growing time in domain I a wave of rarefaction is formed; in domain II, which expands with time, a wave of compression which turns over at great changes in the density A; while in domain III there is a weakly disturbed plasma. Domains I and III are quiet domains, therefore, gas-dynamical plasma equations of type (3.2) are used in them, while in domain II equations (4.21) are solved.

For a numerical realization of the discussed model, the domain $0 \le x \le L$ is divided into K cells of the same length h. In line with the initial condition (4.24), we set the values of the functions n, u_e, B, T in the mesh nodes $x_k = kh$ ($k = 0, 1, ..., K$). In every cell of domain II we place a certain number of 'big' ions in proportion to the density value; for every ion we set $u_j = 0$, $v_j = -m_e/m_i \dfrac{1}{n} \partial B/\partial x$, where j is a number of the particle ($j = 1, 2, ..., J$; J is the total number of the particles). The ions move as described by equations (4.21)–(4.23), the fields E_x, E_y, B values are interpolated into the ions site. By the coordinates and velocities of the 'big' ions trapped in the corresponding cells of the spatial mesh we find the plasma density n_k and the ion macroscopic velocity $\langle u_i \rangle_k \equiv u_k$ in the mesh nodes. The rest of the equations, which are of a gas-dynamical character, are solved throughout the domain $0 \le x \le L$ using difference schemes.

Let us now consider a method of solving equation (4.23) for the magnetic field and temperature under the assumption that both the plasma density n_k and the ion macroscopic velocity u_k in the mesh nodes have already been determined by the positions and velocities of the particles with the procedures described above.

Let us introduce the auxiliary function

$$Q = B - \frac{m_e}{m_i} \frac{\partial}{\partial x}\left(\frac{1}{\varrho} \frac{\partial B}{\partial x} \right), \qquad \varrho = n/n_0,$$

which obeys the equation

$$\frac{\partial Q}{\partial t} + \frac{\partial}{\partial x}(uQ) = (m_i/m_e)\,\mathcal{H}(B - Q), \qquad \mathcal{H} \equiv v_{ef}/\Omega_B.$$

To determine the function Q and temperature T in the mesh nodes we shall use the following difference equations with the sign of the ion macroscopic velocity allowed for:

$$Q_k^{n+1} = Q_k^n - \frac{\tau}{h}\left(u_k\,\Delta Q_k^n + \frac{u_{k+1} - u_{k-1}}{2} Q_k^n \right)$$

$$+ (m_i/m_e)\,(\mathcal{H}\tau/4)\,(B_{k-1}^n + 2B_k^n + B_{k+1}^n - Q_{k-1}^n - 2Q_k^n - Q_{k+1}^n),$$

$$(4.25)$$

$$T_k^{n+1} = T_k^n - \frac{\tau}{h}\left(u_k \Delta T_k^n + \frac{u_{k+1} - u_{k-1}}{3} T_k^n\right)$$

$$+ (\mathscr{H}\tau/3h^2)\varrho_k^{-2}(B_{k+1}^n - B_{k-1}^n)^2 + (\chi\tau/3h^2)\varrho_k^{-1}[(\varrho_{k+1}$$

$$+ \varrho_k)(T_{k+1}^n - T_k^n) - (\varrho_k + \varrho_{k-1})(T_k^n - T_{k-1}^n)], \qquad (4.25)$$

where

$$\Delta f_k = \begin{cases} f_k - f_{k-1} & \text{at} \quad u_k > 0 \\ f_{k+1} - f_k & \text{at} \quad u_k < 0. \end{cases}$$

Then, using the method of sweeping, we find the magnetic field distribution, making use of the obtained values of the function Q_k^{n+1} in the mesh nodes:

$$(\varrho_{k+1} + \varrho_k)^{-1} B_{k+1}^{n+1} - [(\varrho_{k+1} + \varrho_k)^{-1} + (\varrho_k + \varrho_{k+1})^{-1}$$

$$+ (m_i h^2/2m_e)] B_k^{n+1} - (\varrho_{k-1} + \varrho_k)^{-1} B_{k-1}^{n+1} \qquad (4.26)$$

$$= -(m_i h^2/2m_e) Q_k^{n+1}.$$

The best combination of the finite-difference method with the method of particles is ensured by the difference schemes that are sufficiently close to the algorithms of the method of particles in the limiting case of a great number of particles. As has been shown by the analysis of the trial calculations, such a quality is inherent to the schemes wherein the convective transfer is realized with the velocity direction accounted for, i.e. the schemes of type (4.25). A cycle of calculations is terminated by determining B_k^{n+1} by formulae (4.26), then it is repeated until the whole time interval of interest is covered.

Calculations by the combined model demonstrate that at the initial drops in the density and the magnetic field $A < 12$, the evolution of the step (4.24) results in the formation of a moving towards the positive values of the coordinate x laminar shock wave with the subcritical parameters, and a propagating in the opposite direction wave of rarefaction. The shock wave has an oscillatory front conditioned by the dispersion and related to the electron inertia (Fig. 15, where 1 is the magnetic field profile, 2 is the velocity profile). The scale of oscillations is determined by the dispersion length $\delta_e = c/\omega_{0e}$. The dissipative size due to finite conductivity equals $[\mu_0\sigma V_A(M - 1)]^{-1} = \delta_d$;

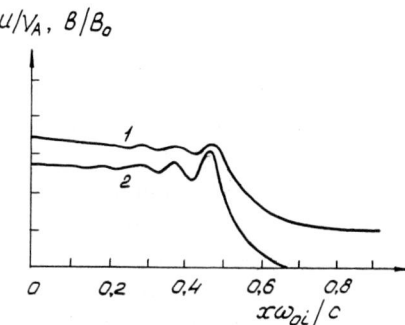

u/v_A, B/B_o

Fig. 15. Shock wave with an oscillatory structure ($A = 10$). 1 — magnetic field profile, 2 — longitudinal plasma velocity profile.

in the described series of calculations $\delta_d = 0.2c/\omega_{oi} = 0.2(m_i/m_e)^{1/2}\delta_e$. The presence of finite conductivity results in a greater width of the magnetic field profile as compared to the profile of the longitudinal velocity and the plasma particle density. There occurs no front over-turning and particle mixing. These data are in full accord with the calculated results based on the two-component gas-dynamical model.

With growing initial drop of density a character of the flow changes. At $12 < A < 25$ and at small times from the beginning of the dis-continuity disintegration, the wave is an unsteady laminar wave with an oscillatory front. With growing time, however, the amplitude of the front oscillation of the profile of the ion (electron) longitudinal velo-city increases up to critical, there occurs a wave overturning followed by the formation of groups of fast ions, reflected by the front, and a precursor on the magnetic field profile. Figure 16 presents the profiles of the longitudinal velocity of ions in the shock wave formed at the discontinuity disintegration with $A = 20$ at the moments of time $t = 0.3\Omega_B^{-1}$ (before overturning) and $t = 0.4\Omega_B^{-1}$ (after overturning), where $\Omega_B = eB_0/m_i$. Subsequent to the particle outburst from the front oscillations, its amplitude reduces, then the amplitude of the next oscillation increases up to critical, then there occurs a particle outburst from it and so on. A decrease in the amplitude of the front oscillation and the formation of fast particles is accompanied by a deceleration of the shock wave. Then the wave velocity increases, there arises a new group of fast particles and so forth. As a result, there arises a regime that can be referred to as pulsating.

At $A \gtrsim 25$ a character of the discontinuity disintegration changes again: the shock wave amplitude always remains supercritical, and the particle outburst occurs continuously. The group of fast particles now comprises both the ions from the undisturbed region, which are reflected by the wave front, and the ions that come with the wave. These two ion groups acquire oppositely directed transversal veloci- ties. On the magnetic field profile a pronounced precursor, differing in amplitude and velocity from the basic front, is formed, with its spatial size determined by the Larmor ion radius. Figure 17 shows the magnetic field profile at the moment of time $t = 3\Omega_B^{-1}$, which is a result of the evolution of the initial disintegration with the density drop $A = 30$. Figure 18 presents in detail the domain of the basic front and the precursor, as well as the position and the longitudinal velocities of ions. In the calculation series discussed, the velocity of the basic front is approximately $6V_A$, the maximum longitudinal ion velocity $8V_A$, is and the precursor velocity $11.5V_A$. Unlike a strong

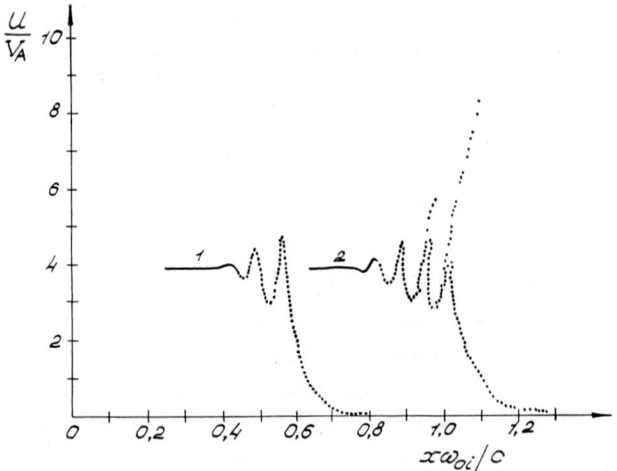

Fig. 16. Profile of ion longitudinal velocity in a shock wave ($A = 20$).
$1 - t = 0.3\Omega_B^{-1}$, $2 - t = 0.4\Omega_B^{-1}$.

shock wave in a non-isothermal plasma with no magnetic field, when the precursor velocity equals the maximum ion velocity u_{max}, in the discussed case with a magnetic field the precursor velocity is less than the maximum longitudinal ion velocity. It is associated with the magnetic field effect resulting in ions turning.

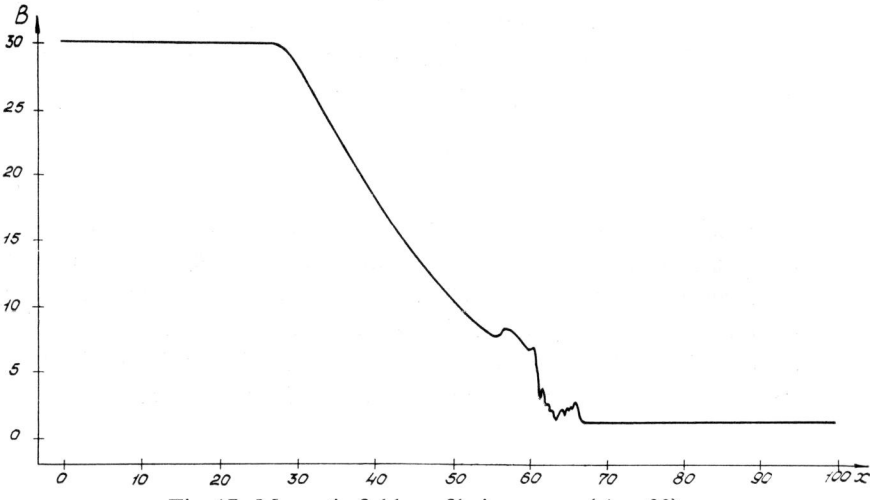

Fig. 17. Magnetic field profile in a wave ($A = 30$).

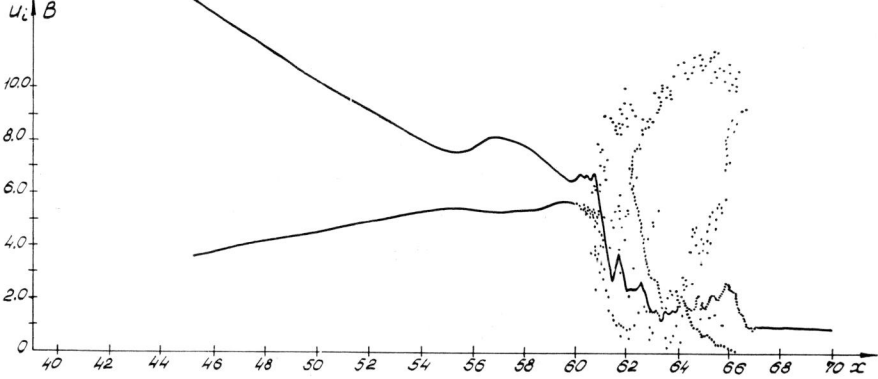

Fig. 18. Magnetic field profile, localization and ion longitudinal velocities in the main front and in the precursor ($A = 30$).

Chapter 5

Two-Dimensional Problems of Magnetogas-dynamics

Let us consider two classes of self-consistent two-dimensional problems of single-liquid magnetogasdynamics: (1) formation and structure of the bow shock waves arising when a supersonic flow of a heat-conducting plasma with a finite conductivity flows around axially-symmetrical blunt bodies; (2) dynamics of plasma flows in the vicinity of neutral lines and magnetic field planes, and of processes of magnetic force lines reconnection.

5.1. Bow Shock Waves

To study the flows with the shock waves arising in the course of blunt convex bodies moving with supersonic speeds in a plasma, it is convenient to make use of the curvilinear 'natural' coordinates connected with the body surface. This local system of coordinates is determined by the unit vectors along the external normal $\mathbf{e}_n$, along the tangential $\mathbf{e}_s$, along the binormal to the body surface $\mathbf{e}_\varphi$. In the coordinates n, s the governing system of two-dimensional equations for a compressible viscous heat-conducting conductive gas can be written in the following dimensionless form:

$$\frac{\partial \varrho}{\partial t} + u \frac{\partial \varrho}{\partial n} + \frac{v}{h_s} \frac{\partial \varrho}{\partial s} + \frac{\varrho}{h_s h_\varphi^k} \left(\frac{\partial}{\partial n} h_s h_\varphi^k u + \frac{\partial}{\partial s} h_\varphi^k v \right) = 0,$$

$$\left(\frac{\partial}{\partial t} + u\frac{\partial}{\partial n} + \frac{v}{h_s}\frac{\partial}{\partial s}\right)u - Kv^2/h_s + \frac{1}{\varrho}\frac{\partial p}{\partial n} + \frac{B}{\varrho h_s M_A^2}\left(\frac{\partial}{\partial n}h_s B\right)$$

$$-\frac{\partial H}{\partial s}\right) = \frac{1}{\varrho Re}\left(\frac{4}{3}\frac{\partial}{\partial n}\mu\frac{\partial u}{\partial n} + \frac{1}{h_s}\frac{\partial}{\partial s}\frac{\mu}{h_s}\frac{\partial u}{\partial s} + F_1\right),$$

$$\left(\frac{\partial}{\partial t} + u\frac{\partial}{\partial n} + \frac{v}{h_s}\frac{\partial}{\partial s}\right)v + Kuv/h_s + \frac{1}{\varrho h_s}\frac{\partial p}{\partial s}$$

$$-\frac{H}{\varrho h_s M_A^2}\left(\frac{\partial}{\partial n}h_s B - \frac{\partial H}{\partial s}\right) = \frac{1}{\varrho Re}\left(\frac{\partial}{\partial n}\mu\frac{\partial v}{\partial n} + \frac{4}{3h_s}\frac{\partial}{\partial s}\frac{\mu}{h_s}\frac{\partial v}{\partial s} + F_2\right),$$

$$\left(\frac{\partial}{\partial t} + u\frac{\partial}{\partial n} + \frac{v}{h_s}\frac{\partial}{\partial s}\right)T + \frac{\gamma p}{\varrho h_s h_\varphi^k}\left(\frac{\partial}{\partial n}h_s h_\varphi^k u + \frac{\partial}{\partial s}h_\varphi^k v\right)$$

$$= \frac{\gamma}{\varrho Re}\left(\frac{\partial}{\partial n}\lambda\frac{\partial T}{\partial n} + \frac{1}{h_s}\frac{\partial}{\partial s}\frac{\lambda}{h_s}\frac{\partial T}{\partial s}\right)$$

$$+ \frac{\gamma}{\varrho h_s \sigma\, Re_m\, M_A^2}\left(\frac{\partial}{\partial n}h_s B - \frac{\partial H}{\partial s}\right)^2 + F_3,$$

$$\frac{\partial H}{\partial t} + \frac{1}{h_s}\left(\frac{u}{h_\varphi^k}\frac{\partial}{\partial n}h_s h_\varphi^k H + \frac{v}{h_\varphi^k}\frac{\partial}{\partial s}h_\varphi^k H + H\frac{\partial v}{\partial s} - B\frac{\partial u}{\partial s}\right)$$

$$+ \frac{1}{h_s h_\varphi^k\, Re_m}\frac{\partial}{\partial s}\frac{h_\varphi^k}{h_s \sigma}\left(\frac{\partial}{\partial n}h_s B - \frac{\partial H}{\partial s}\right) = 0, \qquad (5.1)$$

$$\frac{\partial B}{\partial t} + \frac{u}{h_\varphi^k}\frac{\partial}{\partial n}h_\varphi^k B + \frac{v}{h_s h_\varphi^k}\frac{\partial}{\partial s}h_\varphi^k B - H\frac{\partial v}{\partial n} + B\frac{\partial u}{\partial n}$$

$$+ \frac{KvH}{h_s} - \frac{1}{h_\varphi^k\, Re_m}\frac{\partial}{\partial n}\frac{h_\varphi^k}{h_s \sigma}\left(\frac{\partial}{\partial n}h_s B - \frac{\partial H}{\partial s}\right) = 0,$$

$$p = (\gamma - 1)\varrho T/\gamma;$$

$$F_1 = \frac{1}{h_s}\frac{\partial}{\partial s}\mu\left(\frac{\partial v}{\partial n} - \frac{Kv}{h_s}\right) - \frac{2}{3}\frac{\partial}{\partial n}\mu\left[\frac{1}{h_s}\frac{\partial v}{\partial s} + \frac{Ku}{h_s}\right.$$

$$\left. + k\left(\frac{u\cos\alpha}{h_\varphi} + \frac{v}{h_s h_\varphi}\frac{\partial h_\varphi}{\partial s}\right)\right] + 2\mu\left[\frac{K}{h_s}\left(\frac{\partial u}{\partial n} - \frac{1}{h_s}\frac{\partial v}{\partial s} - \frac{Ku}{h_s}\right)\right.$$

$$+ \frac{k\cos\alpha}{h_\varphi}\left(\frac{\partial u}{\partial n} + \frac{u\cos\alpha}{h_\varphi} + \frac{v}{h_s h_\varphi}\frac{\partial h_\varphi}{\partial s}\right)$$

$$- \frac{k}{2h_s h_\varphi}\left(\frac{1}{h_s}\frac{\partial u}{\partial s} - \frac{Kv}{h_s} + \frac{\partial v}{\partial n}\right)\frac{\partial h_\varphi}{\partial s}\Bigg],$$

$$F_2 = \frac{\partial}{\partial n}\frac{\mu}{h_s}\left(\frac{\partial u}{\partial s} - Kv\right) - \frac{2}{3h_s}\frac{\partial}{\partial s}\mu\left[\frac{\partial u}{\partial n}\right.$$

$$\left. - \frac{2Ku}{h_s} + k\left(\frac{u\cos\alpha}{h_\varphi} + \frac{v}{h_s h_\varphi}\frac{\partial h_\varphi}{\partial s}\right)\right]$$

$$+ \mu\left[\left(\frac{1}{h_s}\frac{\partial u}{\partial s} - \frac{Kv}{h_s} + \frac{\partial v}{\partial n}\right)\left(\frac{2K}{h_s} + \frac{k\cos\alpha}{h_\varphi}\right)\right.$$

$$+ \frac{2k}{h_s h_\varphi}\frac{\partial h_\varphi}{\partial s}\left(\frac{1}{h_s}\frac{\partial v}{\partial s} + \frac{Ku}{h_s} - \frac{u\cos\alpha}{h_\varphi} - \frac{v}{h_s h_\varphi}\frac{\partial h_\varphi}{\partial s}\right)\Bigg],$$

$$F_3 = 2\mu\left\{\frac{1}{h_s^2}\left(\frac{\partial v}{\partial s} + Ku\right)^2 + \left(\frac{\partial u}{\partial n}\right)^2\right.$$

$$+ kh_\varphi^{-2}\left(u\cos\alpha + \frac{v}{h_s}\frac{\partial h_\varphi}{\partial s}\right)^2$$

$$+ \frac{1}{2}\left(\frac{1}{h_s}\frac{\partial u}{\partial s} - \frac{Kv}{h_s} + \frac{\partial v}{\partial n}\right)^2 - \frac{1}{3}\left[\frac{1}{h_s}\left(\frac{\partial v}{\partial s} + Ku\right)\right.$$

$$\left.\left. + k\left(\frac{u\cos\alpha}{h_\varphi} + \frac{v}{h_s h_\varphi}\frac{\partial h_\varphi}{\partial s}\right) + \frac{\partial u}{\partial n}\right]^2\right\}$$

$$+ \lambda\left[\frac{k}{h_s^2 h\varphi}\frac{\partial h\varphi}{\partial s}\frac{\partial T}{\partial s} + \left(\frac{K}{h_s} + \frac{k\cos\alpha}{h_\varphi}\right)\frac{\partial T}{\partial n}\right].$$

The above system of equations was written under the supposition that the vectors of the velocity and the magnetic field strength lie in one plane. The following denotations were introduced: ϱ is the density; u, v are the velocity vector projections onto the axes n, s; p is the pressure; T is the temperature; H, B are the projections of the vector of the magnetic field onto the axes n, s; λ is the heat conductivity coefficient; μ is the coefficient of dynamical viscosity; σ is the conductivity, $k = 0$ corresponds to the plane case, $k = 1$ — to the axissymmetrical case, $R = R(s)$ is the distance from the body axis to its surface; $\alpha = \alpha(s)$ is an angle between the symmetry axis and the tangential to the body contour; $K = K(s)$ is the longitudinal curvature of the body surface; $h_s = 1 + K(s)\,n$, $h_\varphi = R(s) + n\cos\alpha(s)$ are the Lamé coefficients. The density, velocity and strength of the magnetic field are normalized with respect to the values ϱ_∞, u_∞, H_∞, corresponding to the undisturbed flow far ahead of the body, while the temperature is normalized to the quantity u_∞^2/c_p; the radius of the body curvature in the front critical point is chosen as a characteristic length L. Further, $Re = (\varrho u)_\infty L/\mu_\infty$ is the Reynolds number, $Re_m = \mu_0(\sigma u)_\infty L$ is the magnetic Reynolds number, $\gamma = c_p/c_v$, $M_A^2 = u_\infty^2/V_{A\infty}^2$. Functional dependence of the dissipative coefficients μ, λ, σ will be detailed below when describing concrete calculations. A solution to system (5.1) is sought in the domain D, limited by the upper boundary $R_1(s)$, which is chosen by trial calculations, by the intervals of the symmetry axis and by the body surface ($n = 0$, $0 \leq s \leq s_1$). On the symmetry axis we set the conditions

$$\partial\varrho/\partial s = \partial u/\partial s = \partial T/\partial s = v = 0 \qquad (5.2)$$

and on the body surface

$$u(0, s) = v(0, s) = 0,$$

$$\frac{\partial T}{\partial n}(0, s) = 0 \quad \text{or} \quad T(0, s) = T_0. \qquad (5.3)$$

On the part $0 \leq s \leq s_0$ of the external boundary $R_1(s)$ we set an undisturbed flow

$$\varrho = 1, \quad u = H = -\sin\alpha(s), \quad v = B = \cos\alpha(s),$$

$$T = (\gamma - 1)^{-1} M_\infty^{-2}, \quad M_\infty = u_\infty(\gamma p_\infty/\varrho_\infty)^{-1/2} \qquad (5.4)$$

and on the part $s_0 \leq s \leq s_1$ of the external boundary $R_1(s)$ we set the conditions

$$\frac{\partial}{\partial l} (\varrho, u, v, T, H, B) = 0 \tag{5.5}$$

where the direction l coincides with the direction of the symmetry axis, and s_0 is chosen by trial calculations.

To determine a steady solution of problem (5.1)–(5.5) let us make use of the method of adjustment [22, 23], assuming that the solution exists and is unique. Since setting the initial conditions is in this case sufficiently arbitrary, the values corresponding to the undisturbed flow were chosen as the initial sought functions throughout the domain. Let us transform the domain D into a unit square through the formulae

$$\xi = s/\bar{s}, \quad \eta = n/R_1(s), \tag{5.6}$$

where $\bar{s} = s_1$, if the solution is sought throughout the domain, and $\bar{s} = s_0$ in case of a truncated domain. As a result, we get the following system of equations:

$$\frac{\partial f}{\partial t} + \Omega f = F + F_m \tag{5.7}$$

$$\frac{\partial g}{\partial t} + \mathcal{W} g = \Phi, \tag{5.8}$$

where

$$\Omega = \sum_{j=1}^{4} \Omega_j,$$

$$f = \begin{bmatrix} \varrho \\ u \\ v \\ T \end{bmatrix},$$

$$
\boldsymbol{F} = \left\{
\begin{array}{c}
0 \\[2ex]
F_1 - \dfrac{a}{\varrho h_s Re}\left(b\,\dfrac{\partial}{\partial\eta}\,\dfrac{\mu}{h_s}\,\dfrac{\partial}{\partial\xi} + \dfrac{\partial}{\partial\xi}\,\dfrac{\mu b}{h_s}\,\dfrac{\partial}{\partial\eta}\right)u \\[3ex]
F_2 - \dfrac{4a}{3\varrho h_s Re}\left(b\,\dfrac{\partial}{\partial\eta}\,\dfrac{\mu}{h_s}\,\dfrac{\partial}{\partial\xi} + \dfrac{\partial}{\partial\xi}\,\dfrac{\mu b}{h_s}\,\dfrac{\partial}{\partial\eta}\right)v \\[3ex]
F_3 - \dfrac{\gamma a}{\varrho h_s Re}\left(b\,\dfrac{\partial}{\partial\eta}\,\dfrac{\lambda}{h_s}\,\dfrac{\partial}{\partial\xi} + \dfrac{\partial}{\partial\xi}\,\dfrac{\lambda b}{h_s}\,\dfrac{\partial}{\partial\eta}\right)T
\end{array}
\right\},
$$

$$
\boldsymbol{\Omega_1 f} = \left\{
\begin{array}{c}
\dfrac{av}{h_s}\,\dfrac{\partial\varrho}{\partial\xi} \\[3ex]
\dfrac{av}{h_s}\,\dfrac{\partial u}{\partial\xi} - \dfrac{a^2}{\varrho h_s Re}\,\dfrac{\partial}{\partial\xi}\,\dfrac{\mu}{h_s}\,\dfrac{\partial u}{\partial\xi} \\[3ex]
\dfrac{av}{h_s}\,\dfrac{\partial v}{\partial\xi} - \dfrac{4a^2}{3\varrho h_s Re}\,\dfrac{\partial}{\partial\xi}\,\dfrac{\mu}{h_s}\,\dfrac{\partial v}{\partial\xi} \\[3ex]
\dfrac{av}{h_s}\,\dfrac{\partial T}{\partial\xi} - \dfrac{\gamma a^2}{\varrho h_s Re}\,\dfrac{\partial}{\partial\xi}\,\dfrac{\lambda}{h_s}\,\dfrac{\partial T}{\partial\xi}
\end{array}
\right\},
$$

$$
\boldsymbol{\Omega_2 f} = \left\{
\begin{array}{c}
\left(\dfrac{u}{R_1} - \dfrac{bv}{h_s}\right)\dfrac{\partial\varrho}{\partial\eta} \\[3ex]
\left(\dfrac{u}{R_1} - \dfrac{bv}{h_s}\right)\dfrac{\partial u}{\partial\eta} - \dfrac{1}{\varrho Re}\left(\dfrac{4}{3R_1^2}\,\dfrac{\partial}{\partial\eta}\,\mu + \dfrac{b}{h_s}\,\dfrac{\partial}{\partial\eta}\,\dfrac{\mu b}{h_s}\right)\dfrac{\partial u}{\partial\eta} \\[3ex]
\left(\dfrac{u}{R_1} - \dfrac{bv}{h_s}\right)\dfrac{\partial v}{\partial\eta} - \dfrac{1}{\varrho Re}\left(\dfrac{4b}{3h_s}\,\dfrac{\partial}{\partial\eta}\,\dfrac{\mu b}{h_s} + \dfrac{1}{R_1^2}\,\dfrac{\partial}{\partial\eta}\,\mu\right)\dfrac{\partial v}{\partial\eta} \\[3ex]
\left(\dfrac{u}{R_1} - \dfrac{bv}{h_s}\right)\dfrac{\partial T}{\partial\eta} - \dfrac{\gamma}{\varrho Re}\left(\dfrac{b}{h_s}\,\dfrac{\partial}{\partial\eta}\,\dfrac{\lambda b}{h_s} + \dfrac{1}{R_1^2}\,\dfrac{\partial}{\partial\eta}\,\lambda\right)\dfrac{\partial T}{\partial\eta}
\end{array}
\right\},
$$

$$\Omega_3 f = \left\{ \begin{array}{c} \dfrac{a\varrho}{h_s h_\varphi^k} \dfrac{\partial}{\partial \xi} h_\varphi^k v \\[3mm] -Kv^2 h_s^{-1} \\[3mm] Kuv h_s^{-1} + (\gamma - 1)(\gamma h_s)^{-1}\left(a\dfrac{\partial T}{\partial \xi} + \dfrac{T}{\varrho}\dfrac{\partial \varrho}{\partial \xi}\right) \\[3mm] \dfrac{(\gamma - 1)\, aT}{h_s h_\varphi^k} \dfrac{\partial}{\partial \xi} h_\varphi^k v \end{array} \right\}$$

$$\Omega_4 f = \left\{ \begin{array}{c} \dfrac{\varrho}{h_s h_\varphi^k}\left(\dfrac{1}{R_1}\dfrac{\partial}{\partial \eta} h_s h_\varphi^k u - b\dfrac{\partial}{\partial \eta} h_\varphi^k v\right) \\[3mm] \dfrac{\gamma - 1}{\gamma R_1}\left(\dfrac{T}{\varrho}\dfrac{\partial \varrho}{\partial \eta} + \dfrac{\partial T}{\partial \eta}\right) \\[3mm] -b\dfrac{\gamma - 1}{\gamma h_s}\left(\dfrac{\partial T}{\partial \eta} + \dfrac{T}{\varrho}\dfrac{\partial \varrho}{\partial \eta}\right) \\[3mm] (\gamma - 1)\, T h_s^{-1} h_\varphi^{-k}\left(\dfrac{1}{R_1}\dfrac{\partial}{\partial \eta} h_s h_\varphi^k u - b\dfrac{\partial}{\partial \eta} h_\varphi^k v\right) \end{array} \right\}$$

$$F_m = \left\{ \begin{array}{c} 0 \\[3mm] -B(\varrho h_s M_A^2)^{-1}\left(\dfrac{\partial}{\partial n} h_s B - \dfrac{\partial H}{\partial s}\right) \\[3mm] H(\varrho h_s M_A^2)^{-1}\left(\dfrac{\partial}{\partial n} h_s B - \dfrac{\partial H}{\partial s}\right) \\[3mm] \gamma(\varrho h_s^2 \sigma\, Re_m\, M_A^2)^{-1}\left(\dfrac{\partial}{\partial n} h_s B - \dfrac{\partial H}{\partial s}\right)^2 \end{array} \right\}$$

The operators Ω_1, Ω_2 allow for the convective and viscous terms, and Ω_3, Ω_4 for the pressure in the equations of motion and div u in other equations.

$$g = \begin{pmatrix} H \\ B \end{pmatrix}, \qquad \mathscr{W} = \sum_{j=1}^{2} \mathscr{W}_j,$$

$$\mathscr{W}_1 = \begin{bmatrix} \dfrac{av}{h_s h_\varphi^k} \dfrac{\partial}{\partial \xi} h_\varphi^k - \dfrac{a^2}{h_s h_\varphi^k Re_m} \dfrac{\partial}{\partial \xi} \dfrac{h_\varphi^k}{h_s \sigma} \dfrac{\partial}{\partial \xi}, & 0 \\[3mm] 0, & \dfrac{av}{h_s h_\varphi^k} \dfrac{\partial}{\partial \xi} h_\varphi^k \end{bmatrix},$$

$$\mathscr{W}_2 = \begin{bmatrix} h_\varphi^{-k}\left(\dfrac{u}{R_1} - \dfrac{bv}{h_s}\right)\dfrac{\partial}{\partial \eta} h_\varphi^k - \dfrac{b}{h_s h_\varphi^k Re_m}\dfrac{\partial}{\partial \eta} \dfrac{h_\varphi^k b}{h_s \sigma}\dfrac{\partial}{\partial \eta}, \\[3mm] -\dfrac{b}{h_s h_\varphi^k R_1 Re_m}\dfrac{\partial}{\partial \eta}\dfrac{h_\varphi^k}{h_s \sigma}\dfrac{\partial}{\partial \eta} h_s \\[3mm] -(h_\varphi^k R_1 Re_m)^{-1}\dfrac{\partial}{\partial \eta}\dfrac{h_\varphi^k b}{h_s \sigma}\dfrac{\partial}{\partial \eta}, \\[3mm] h_\varphi^{-k}\left(\dfrac{u}{R_1} - \dfrac{bv}{h_s}\right)\dfrac{\partial}{\partial \eta} h_\varphi^k - (h_\varphi^k R_1 Re_m)^{-1}\dfrac{\partial}{\partial \eta}\dfrac{h_\varphi^k}{h_s \sigma}\dfrac{\partial}{\partial \eta} h_s \end{bmatrix},$$

$$\Phi = \begin{bmatrix} -h_s^{-1}\left[KuH + \left(a\dfrac{\partial v}{\partial \xi} - b\dfrac{\partial v}{\partial \eta}\right)H - \left(a\dfrac{\partial u}{\partial \xi} - b\dfrac{\partial u}{\partial \eta}\right)B\right] \\[3mm] +\dfrac{a}{h_s h_\varphi^k Re_m}\left[\dfrac{\partial}{\partial \xi}\dfrac{h_\varphi^k}{h_s \sigma}\left(\dfrac{1}{R_1}\dfrac{\partial}{\partial \eta} h_s B + b\dfrac{\partial H}{\partial \eta}\right) + b\dfrac{\partial}{\partial \eta}\dfrac{h_\varphi^k}{h_s \sigma}\dfrac{\partial H}{\partial \xi}\right], \\[3mm] -h_s^{-1}KvH - R_1^{-1}\left(\dfrac{\partial u}{\partial \eta}B - \dfrac{\partial v}{\partial \eta}H\right) + \dfrac{a}{h_\varphi^k R_1 Re_m}\dfrac{\partial}{\partial \eta}\dfrac{h_\varphi^k}{h_s \sigma}\dfrac{\partial H}{\partial \xi} \end{bmatrix},$$

In these expressions $a = \bar{s}^{-1}$, $b = \eta R_1'/R_1$.

In the transformed domain $0 \le \xi \le 1$, $0 \le \eta \le 1$ let us introduce a difference mesh with the steps $h_1 = 1/I$, $h_2 = 1/J$, where I, J are the number of the nodes along the directions ξ and η, respectively; and let τ be an iteration parameter (or a step in time in an unsteady case). Besides, let us introduce the mesh vector-functions f_h^n, g_h^n, F_h^n, Φ_h^n

coinciding with the functions f, g, F, Φ in the mesh nodes (i, j, n) and the difference operators Ω_j^h, $\mathscr{W}_j^h$, which approximate the initial differential operators Ω_j, $\mathscr{W}_j$ to the second degree of accuracy. The convective terms in the operators Ω_1^h, Ω_2^h, $\mathscr{W}_1^h$, $\mathscr{W}_2^h$ are approximated by the formulae

$$(c\Lambda_1 f)_{i,j} = [(c + |c|)/4h_1]\,(3f_{i,j} - 4f_{i-1,j} + f_{i-2,j}) \qquad (5.9)$$
$$+ [(c - |c|)/4h_1]\,(4f_{i+1,j} - 3f_{i,j} - f_{i+2,j}),$$
$$(c\Lambda_2 f)_{i,j} = [(c + |c|)/4h_2]\,(3f_{i,j} - 4f_{i,j-1} - f_{i,j-2})$$
$$+ [(c - |c|)/4h_2]\,(4f_{i,j+1} - 3f_{i,j} - f_{i,j+2}).$$

The viscous terms in the operators

$$\Omega_1^h, \ \Omega_2^h, \ \mathscr{W}_1^h, \ \mathscr{W}_2^h$$

are approximated by the formulae

$$(\Lambda_1 a \Lambda_1 f)_{i,j} = (2h_1^2)^{-1}\,[a_{i+1,j} + a_{i,j})\,(f_{i+1,j} - f_{i,j})$$
$$- (a_{i,j} + a_{i-1,j})\,(f_{i,j} - f_{i-1,j})],$$
$$(\Lambda_2 b \Lambda_2 f)_{i,j} = (2h_2^2)^{-1}\,[(b_{i,j+1} + b_{i,j})\,(f_{i,j+1} - f_{i,j})$$
$$- (b_{i,j} + b_{i,j-1})\,(f_{i,j} - f_{i,j-1}).$$

The mixed derivatives are approximated through the formulae

$$(\Lambda_1 d \Lambda_2 f)_{i,j} = (4h_1 h_2)^{-1}\,[d_{i+1,j}(f_{i+1,j+1} - f_{i+1,j-1})$$
$$- d_{i-1,j}(f_{i-1,j+1} - f_{i-1,j-1})].$$

The first derivatives in the operators Ω_3^h, Ω_4^h are approximated in line with the following rule: if the coefficient c is greater than zero (see (5.9)), then these terms in the equations of continuity and temperature are as follows

$$(c\Lambda_1^{(-1)} f)_{i,j} = (c/2h_1)\,(3f_{i,j} - 4f_{i-1,j} - f_{i-2,j})$$

and in the equations of motion

$$(c\Lambda_1^{(+)} f)_{i,j} = (c/2h_1)\,(4f_{i+1,j} - 3f_{i,j} - f_{i+2,j})$$

and vice versa in the case when $c < 0$.

Besides the operators Ω^h_j, $\mathscr{W}^h_j$ let us also introduce the difference operators $\bar{\Omega}^h_j$, $\bar{\mathscr{W}}^h_j$, wherein the first derivatives are approximated to the first order by the formulae

$$(c\bar{A}_1 f)_{i,j} = [(c + |c|)/2h_1](f_{i,j} - f_{i-1,j}) + [(c - |c|)/2h_1](f_{i+1,j} - f_{i,j}),$$

$$(\bar{A}_1^{(+)}f)_{i,j} = (f_{i+1,j} - f_{i,j})/h_1, \quad (\bar{A}_1^{(-)}f)_{i,j} = (f_{i,j} - f_{i-1,j})/h_1.$$

The whole problem (5.7), (5.8) is solved in two steps: first from system (5.7) we determine the values of the gas-dynamical functions ϱ, u, v, T by the known distribution of the magnetic field; then from system (5.8) we find the new values of the magnetic field strength.

To solve the system of equations (5.7) we make use of the following scheme of fractional steps:

$$(E + \tau\bar{\Omega}^h_1)\varphi^{n+1/4} = -\tau(\Omega^h f^n_h - F^n_h - F^n_{mh}),$$

$$(E + \tau\bar{\Omega}^h_2)\varphi^{n+2/4} = \varphi^{n+1/4},$$

$$(E + \tau\bar{\Omega}^h_3)\varphi^{n+3/4} = \varphi^{n+2/4},$$

$$(E + \tau\bar{\Omega}^h_4)\varphi^{n+1} = \varphi^{n+3/4},$$

$$f^{n+1}_h = f^n_h + \varphi^{n+1}.$$

(5.10)

Subsequent to excluding the fractional steps, system (5.10) can be written as a scheme of the universal algorithm, having the property of complete approximation:

$$\bar{C}(f^{n+1}_h - f^n_h)/\tau = -\Omega f^n_h + F^n_h + F^n_{mh}$$
(5.11)

where the stabilizing operator $\bar{C} = \prod_{j=1}^{4}(E + \tau\bar{\Omega}^h_j)$.

In the case of a steady problem scheme (5.11) approximates equations (5.7) to the second order of accuracy $O(h^2)$ and is realized through economical three-point sweepings. If we are interested in the solution of the unsteady problem, then the scheme in question approximates the governing differential equations (5.7) to the first order $O(\tau, h)$. In order to get the order of approximation $O(\tau, h^2)$ in case of an unsteady problem, the operator $\bar{C}$ in scheme (5.11) should be replaced with the operator $C = \prod\limits_{j=1}^{4} (E + \tau\Omega_j^h)$; in this case the scheme is realized by the five-point sweepings.

To solve the system of equations (5.8) the following scheme of fractional steps is employed:

$$(E + \tau\overline{\mathcal{W}}_1^h)\,\boldsymbol{\psi}^{n+1/2} = -\tau(\mathcal{W}^h g_h^n - \boldsymbol{\Phi}_h^n),$$

$$(E + \tau\overline{\mathcal{W}}_2^h)\,\boldsymbol{\psi}^{n+1} = \boldsymbol{\psi}^{n+1/2},$$

$$g_h^{n+1} = g_h^n + \boldsymbol{\psi}_h^{n+1}. \tag{5.12}$$

Subsequent to excluding the fractional step we go over to the scheme of the universal algorithm possessing the property of complete approximation:

$$\overline{B}(g_h^{n+1} - g_h^n)/\tau = -\mathcal{W}g_h^n + \boldsymbol{\Phi}_h^n,$$

where

$$\overline{B} = \prod\limits_{j=1}^{2} (E + \tau\overline{\mathcal{W}}_j^h).$$

It follows from the linear analysis given in [23] that scheme (5.10) is absolutely stable in the limiting cases of small and great Reynolds numbers (in the former case one can neglect the inertial terms and pressure forces, in the latter, the viscous terms). Trial calculations have demonstrated that scheme (5.10), is stable at least up to the Courant numbers $K = (\tau/h_1 h_2)\,(h_1\,|v| + h_2\,|u| + c \cdot \max\,(h_1, h_2)) \leq 10$. According to linear analysis, scheme (5.12) is also absolutely stable. Nevertheless, the total scheme for calculating problem (5.7), (5.8) in the two above steps can be conditionally stable in case when the magnetic pressure is great as compared to the gas-kinetic one.

The difference scheme (5.10)–(5.12) has been used to calculate flows of a neutral gas and plasma around a sphere (the axis-symmetrical case) and a cylinder (the plane case). The steady problem is solved by the method of adjustment until $\max\limits_{i,j}|\varrho^{-1}\,\partial\varrho/\partial t| < 10^{-3}$. To get a steady solution in all the considered cases, 150–300 iterations were necessary. The iteration parameter at the beginning of calculations was chosen $\tau = \dfrac{1}{2} \min (h_1, h_2)$, during the next 50–70 steps it increased up to $\tau \approx 3 \max (h_1, h_2)$, then the calculations were carried on with a constant iteration parameter. Changes in τ made it possible to avoid strong disturbances in the vicinity of the body at small times, when the difference between the conditions on the body and the flow, which is close to the undisturbed flow set at $t = 0$, is especially pronounced. An accuracy control was carried out by way of changing the mesh steps (i.e. by changing the position of the upper boundary $R_1(s)$) and by way of checking the laws of mass and energy conservation.

To study a bow shock wave there has been carried out a series of calculations of flows of a neutral gas ($M_A = \infty$ corresponds to this case) and plasma around a sphere which, for simplicity, is considered infinitely conductive and heat-isolated; the boundary condition $H = 0$ is set on its surface. These calculations were carried out at the following parameters: $M_\infty = 2$, $Re = 60$, $Re_m = 100$, $M_A^2 = \infty$; 10; 5; 3; 2.5. Figure 19 presents the position of a shock wave arising at a super-

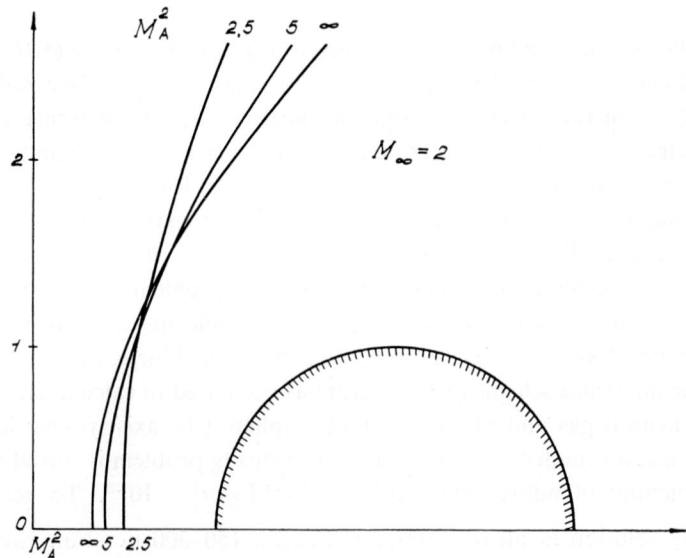

Fig. 19. Shock wave localization in a supersonic flow around a sphere.

sonic flowing around a sphere, which is fixed with respect to the maximum density gradient $|\partial \varrho/\partial n|$. The dependence of a shock wave withdrawal $\bar{x}$ on the Alfven–Mach number M_A at the angles $\theta = 0$, 30, 60° is shown in Fig. 20.

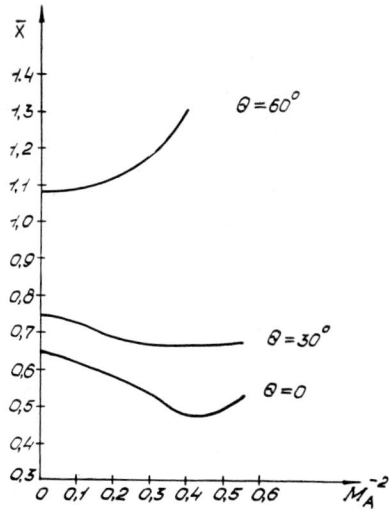

Fig. 20. Dependence of a shock wave retreat on the Mach number M_A at various angles θ read off the stagnation line.

The angle θ is read off from the stagnation line in the clockwise direction. As has been shown by calculations, at $0 \le \theta \le 30°$ a decrease in M_A (which corresponds to an increase in the magnetic field H_∞ at a constant velocity of the on-coming flow u_∞) results in the approaching of the shock wave to the body. At the stagnation line ($\theta = 0$) the withdrawal reaches its minimum value $\bar{x} \approx 0.48$ at $M_A = 1.6$, then a further decrease in M_A is accompanied by growing with drawal. It is related to the fact that a decrease in the Alfven–Mach number at a constant Mach number means a decrease in the number $M = u_\infty(c_\infty^2 + V_{A\infty}^2)^{-1/2} = (M_\infty^{-2} + M_A^{-2})^{-1/2}$ and at $M_A^2 > (1 - M_\infty^{-2})^{-1}$ the whole flow becomes subsonic with respect to the total speed of sound. At large angles ($\theta \gtrsim 60°$) a decrease in the Alfven–

Mach number results in an increase of the shock wave withdrawal from the sphere.

Figure 21 presents distributions of the flow density and longitudinal velocity at the stagnation line for the values $M_A^2 = \infty(1)$, 10 (2), 5 (3), 2.5 (4). The width of the density profile is inversely proportional to

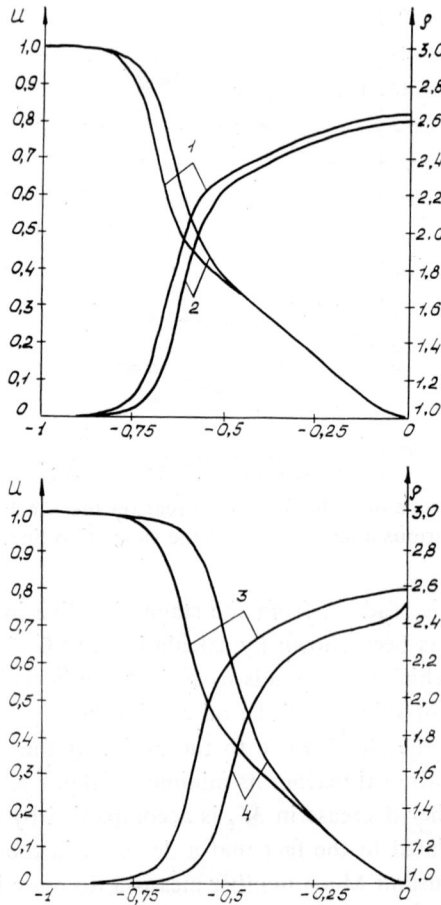

Fig. 21. Distribution of the flow density and longitudinal velocity on the stagnation line.
$1 - M_A^2 = \infty$; $2 - M_A^2 = 10$; $3 - M_A^2 = 5$; $4 - M_A^2 = 2.5$.

the Mach numbers M_∞ and the Reynolds numbers Re and in the case in question, equals $\Delta\varrho \approx 0.2$. In a neutral gas flow the longitudinal velocity profile at the stagnation line has an upward curvature. In case of plasma at $M_A^2 \geq 3$ the longitudinal velocity profile is qualitatively similar to the gas-dynamical one (Fig. 21, line 2), and at a further decrease in the Alfven–Mach number it becomes downward convex near the body (Fig. 21, line 4).

Calculations demonstrate that a magnetic field exerts a stagnating effect on the flow and increases the boundary layer thickness on the body which manifests itself in a decrease of the resistivity coefficient; an analogous effect being valid also for an incompressible conducting fluid. In line with the calculations carried out on flowing around a sphere at various Reynolds numbers (100–1000) an increase in Re_m (i.e. an increase in plasma conductivity with all the parameters constant) results in growing steepness of the magnetic field profile at the stagnation line in front of the body and in reducing thickness of the boundary layer on the body surface.

The analogous results in a qualitative respect have been obtained for a plane flowing around the cylinder, but in this case the shock wave goes from the cylinder at a greater distance than from the sphere.

5.2. Reconnection Processes

Modelling reconnection processes of the lines of force of magnetic fields is of great importance in studying a number of physical phenomena: sun outbursts, changes of the magnetic field topology in the Earth's magnetosphere, plasma instability in thermo-nuclear devices of the tokamak type. In [101] the author studied a two-dimensional MHD plasma flow with a finite conductivity in the plane x, y which is orthogonal to the zero line of the magnetic field. As an initial condition the equilibrium $v - 0$, $\varrho = $ constant was chosen in a hyperbolic magnetic field compressed along the y-axis:

$$B_x^0 = 2^{1/2}q^2y(1 + q^2)^{-1/2}, \quad B_y^0 = 2^{1/2}x(1 + q^2)^{-1/2}, \quad B_z^0 = 0,$$

$$q > 1.$$

The initial plasma pressure is chosen from the condition of plasma equilibrium in a magnetic field and equals $p(x, y, 0) = (q^2 - 1) (x^2 - q^2 y^2) (1 + q^2)^{-1}$. The plasma is disturbed from the equilibrium state by setting its velocity at the upper boundary of the calculation domain. A similar problem was considered in [37] but as an initial condition the plasma equilibrium was taken $v = 0$, $\varrho = $ const. in a hyperbolic magnetic field $B_x^0 = -ky$, $B_y^0 = -kx$, $B_z^0 = 0$, where k is a fixed value of the magnetic field gradient. Such configuration is referred to as the X-type configuration and forceless since $j = \mu_0^{-1}$ rot $B = 0$. The plasma motion is initiated by an electric field which switches on at the initial moment of time at the boundary of the calculation domain. In [105] a neutral uniform current layer is located at the moment $t = 0$ in the magnetic field

$$B_y^0 = B_z^0 = 0, \qquad B_x^0 = \begin{cases} ky, \ |y| \leq L, \\ B_0, \ y > L, \\ -B_0, \ y < -L, \end{cases}$$

where $2L$ is the initial width of the current layer which is disturbed from equilibrium by a local increase in the plasma resistivity. The resistivity distribution then remains constant throughout the calculations.

In [15] the authors numerically solved the problems on plasma flows in the vicinity of a neutral (or zero) surface of the magnetic field and on forming a closed configuration in an open trap under the conditions close to those of laboratory experiments [48]. A zero or a neutral surface is a surface on either side of which the magnetic field is of an opposite direction. To describe the plasma behaviour the following system of equations is used:

$$\frac{\partial \varrho}{\partial t} + \text{div } u = 0,$$

$$\varrho \left(\frac{\partial u}{\partial t} + (u \nabla) u \right) = -\nabla p + [\text{rot } B \cdot B],$$

(5.13)

$$\frac{\partial \boldsymbol{B}}{\partial t} = \text{rot}\,\{[\boldsymbol{u}\boldsymbol{B}] - (\mu_0 \sigma)^{-1}\,\text{rot}\,\boldsymbol{B}\}, \tag{5.13}$$

$$\frac{\partial p}{\partial t} = -\boldsymbol{u}\,\nabla p - \gamma p\,\text{div}\,\boldsymbol{u} + (\gamma - 1)\,\mu_0^{-2}\sigma^{-1}\,(\text{rot}\,\boldsymbol{B})^2 .$$

Since a two-dimensional problem is considered, instead of the magnetic field strength let us introduce a vector potential according to the general definition $\boldsymbol{B} = \text{rot}\,A$. Such a transition is convenient for the following reasons: only one component of the vector potential is other than zero (A_φ in a cylindrical case, A_z in a plane case); the magnetic field components are determined by differentiating this function with respect to the coordinates; the lines $rA_\varphi(A_z) = \text{const.}$ are the lines of force of the magnetic field through which we can easily demonstrate changes in its topology; the plasma conductivity σ is in the equation of induction not under the sign of differentiation but is used as a coefficient.

Taking into account all the above, let us write the governing equations (5.13) in a dimensionless form, having chosen as scales of length, density, magnetic potential, velocity and pressure the following characteristic values: R, ϱ_0, $B_0 R$, $V_A = B_0(\mu_0\varrho_0)^{-1/2}$, $B_0^2/2\mu_0$, respectively (here R is either a width of the plane layer or a radius of the cylindrical chamber).

The plane case:

$$\frac{\partial \varrho}{\partial t} = -u\,\frac{\partial \varrho}{\partial x} - v\,\frac{\partial \varrho}{\partial y} - \varrho\left(\frac{\partial u}{\partial x} + \frac{\partial v}{\partial y}\right),$$

$$\frac{\partial u}{\partial t} = -u\,\frac{\partial u}{\partial x} - v\,\frac{\partial u}{\partial y} - \frac{1}{2\varrho}\,\frac{\partial p}{\partial x} - \frac{1}{\varrho}\,\frac{\partial A}{\partial x}\,\Delta A,$$

$$\frac{\partial v}{\partial t} = -u\,\frac{\partial v}{\partial x} - v\,\frac{\partial v}{\partial y} - \frac{1}{2\varrho}\,\frac{\partial p}{\partial y} - \frac{1}{\varrho}\,\frac{\partial A}{\partial y}\,\Delta A, \tag{5.14a}$$

$$\frac{\partial A}{\partial t} = -u\,\frac{\partial A}{\partial x} - v\,\frac{\partial A}{\partial y} + v_m\,\Delta A,$$

$$\frac{\partial p}{\partial t} = -u\,\frac{\partial p}{\partial x} - v\,\frac{\partial p}{\partial y} + 2(\gamma - 1)\,v_m(\Delta A)^2 - \gamma p\left(\frac{\partial u}{\partial x} + \frac{\partial v}{\partial y}\right),$$

$$\Delta A \equiv \partial^2 A/\partial x^2 + \partial^2 A/\partial y^2 .$$

The cylindrical case:

$$\frac{\partial \varrho}{\partial t} = - u \frac{\partial \varrho}{\partial r} - v \frac{\partial \varrho}{\partial z} - \varrho \left(\frac{\partial u}{\partial r} + \frac{\partial v}{\partial z} + \frac{u}{r} \right),$$

$$\frac{\partial u}{\partial t} = - u \frac{\partial u}{\partial r} - v \frac{\partial u}{\partial z} - \frac{1}{2\varrho} \frac{\partial p}{\partial r} - \frac{1}{\varrho r} \frac{\partial}{\partial r} (rA) \cdot \tilde{\Delta}A,$$

$$\frac{\partial v}{\partial t} = - u \frac{\partial v}{\partial r} - v \frac{\partial v}{\partial z} - \frac{1}{2\varrho} \frac{\partial p}{\partial z} - \frac{1}{\varrho} \frac{\partial A}{\partial z} \tilde{\Delta}A, \qquad (5.14b)$$

$$\frac{\partial A}{\partial t} = - \frac{u}{r} \frac{\partial}{\partial r} (rA) - v \frac{\partial A}{\partial z} + v_m \left(\frac{\partial}{\partial r} \frac{1}{r} \frac{\partial}{\partial r} rA + \frac{\partial^2 A}{\partial z^2} \right),$$

$$\frac{\partial p}{\partial t} = - u \frac{\partial p}{\partial r} - v \frac{\partial p}{\partial z} - \gamma p \left(\frac{1}{r} \frac{\partial}{\partial r} ru + \frac{\partial v}{\partial z} \right) + 2(\gamma - 1) v_m (\tilde{\Delta}A)^2,$$

$$\tilde{\Delta}A \equiv \frac{\partial}{\partial r} \frac{1}{r} \frac{\partial}{\partial r} rA + \frac{\partial^2 A}{\partial z^2}.$$

In equations (5.14) the quantity $v_m = (\mu_0 \sigma)^{-1}$ is a magnetic viscosity; the conductivity is $\sigma = \varrho e^2 / m_i m_e v$; v is the frequency of collisions among the particles, which has been used in calculations as a sum of the Coulomb and anomalous frequencies, i.e.

$$v = v_1 T^{-3/2} + v_a; \qquad v_1, v_a = \text{const}.$$

The undisturbed plasma $(u = v = 0)$ occupies the domain $0 \leq x \leq x_{\max}$, $0 \leq y \leq y_{\max}$ (the plane case), $0 \leq r \leq 1$, $0 \leq z \leq z_{\max}$ (the cylindrical case) and is placed in the magnetic field

$$\boldsymbol{B} = \{th\alpha(y - y_0), 0, 0\} \quad \text{the plane problem},$$
$$\boldsymbol{B} = \{0, 0, th\alpha (r - r_0)\} \quad \text{the cylindrical problem}. \tag{5.15}$$

The coefficient α defines the magnetic field gradient, the coordinates y_0, r_0 outline the position of the zero plane or cylindrical surface on either side of which the magnetic field has an opposite direction.

The only component other than zero of the vector potential corresponding to the distribution of the magnetic field (5.15) in the plane case is determined analytically

$$A(x, y) = \alpha^{-1} \ln ch\alpha(y - y_0)$$

and in the cylindrical case — through the numerical integration,

$$A(r, z) = r^{-1} \int_0^r r' B_z(r', z) \, dr'.$$

For the plane problem the initial plasma density distribution is chosen, in line with [55] as

$$\varrho(x, y, 0) = 1 + Nch^{-2}\alpha(y - y_0), \qquad (5.16a)$$

and for the cylindrical problem, according to the experimental data [48], as

$$\varrho(r, z, 0) = \begin{cases} \max (1, Nch^{-2}\alpha(r - r_0), & r < r_0, \\ \max (0.1, Nch^{-2} \alpha (r - r_0), & r \geq r_0. \end{cases} \qquad (5.16b)$$

Spatial distribution of the plasma pressure at the moment of time $t = 0$ is found out from the condition of plasma equilibrium in a magnetic field: $p + B^2/2\mu_0 = $ const.

Initiation of plasma motion in the plane case is realized through local changes (increase or decrease) in pressure at the initial moment of time

$$p(x, y, 0) = p^0(x, y) + C \exp \{-\beta[(x - x_0)^2 + (y - y_0)^2]\}, \qquad (5.16c)$$

where C is the amplitude of the disturbance, β is its spatial 'width', x_0, y_0 are the disturbance centre coordinates; while in the cylindrical case, by setting the potential disturbance at the boundary of the calculation domain

$$A(1, z, t) = A_0 + A_1 \exp (-\beta_1 z^2) \sin \omega t, \qquad (5.17a)$$

where A_0 is the potential value at $t = 0$, A_1, ω, β are the constants determining the amplitude, frequency and spatial 'width' of the disturbance, respectively.

In the plane case all the boundaries of the calculation domain are chosen in such a way that they are the symmetry planes:

$$0 = v = \frac{\partial u}{\partial y} = \frac{\partial p}{\partial y} = \frac{\partial \varrho}{\partial y} = \frac{\partial^2 A}{\partial y^2} \quad \text{at} \quad y = 0, \; y = y_{max},$$

(5.17b)

$$0 = u = \frac{\partial v}{\partial x} = \frac{\partial p}{\partial x} = \frac{\partial \varrho}{\partial x} = \frac{\partial^2 A}{\partial x^2} \quad \text{at} \quad x = 0, \; x = x_{max}.$$

In the cylindrical case all the boundaries of the calculation domain are the symmetry lines or planes, except for the external surface $r = 1$, corresponding to the radial size of the experimental device. At the above boundaries we have:

$$u = A = \frac{\partial v}{\partial z} = \frac{\partial p}{\partial z} = \frac{\partial \varrho}{\partial z} = 0 \quad \text{at} \quad r = 0,$$

(5.17c)

$$v = \frac{\partial u}{\partial r} = \frac{\partial p}{\partial r} = \frac{\partial \varrho}{\partial r} = \frac{\partial A}{\partial r} = 0 \quad \text{at} \quad z = 0, \; z = z_{max}.$$

Thus, equation (5.14) with the initial data (5.15), (5.16) and the boundary conditions (5.17) are a mathematical formulation of the problem under discussion.

To solve the problem numerically, the following finite-difference scheme is used (for simplicity written for the plane case):

$$u^{n+1} = u - \tau\{u\Lambda_1 u + v\Lambda_2 v + (4\varrho h_1)^{-1} (p_{i+1,j} - p_{i-1,j})$$
$$+ (2\varrho h_1)^{-1} (A_{i+1,j} - A_{i-1,j}) \Lambda A\},$$

$$v^{n+1} = v - \tau\{u\Lambda_1 v + v\Lambda_2 v + (4\varrho h_2)^{-1} (p_{i,j+1} - p_{i,j-1})$$
$$+ (2\varrho h_2)^{-1} (A_{i,j+1} - A_{i,j-1}) \Lambda A\},$$

$$\varrho^{n+1} = \varrho - \tau\{u\Lambda_1 \varrho + v\Lambda_2 \varrho$$
$$+ \varrho[(2h_1)^{-1} (u_{i+1,j}^{n+1} - u_{i-1,j}^{n+1}) + (2h_2)^{-1} (v_{i,j+1}^{n+1} - v_{i,j-1}^{n+1})]\},$$

$$A^{n+1} = A - \tau(u\Lambda_1 A + v\Lambda_2 A - \nu_m \Lambda A),$$

$$p^{n+1} = p - \tau\{u\Lambda_1 p + v\Lambda_2 p + \gamma p[(2h_1)^{-1} (u_{i+1,j}^{n+1} - u_{i-1,j}^{n+1})$$
$$+ (2h_2)^{-1} (v_{i,j+1}^{n+1} - v_{i,j-1}^{n+1})] - 2(\gamma - 1) \nu_m (\Lambda A)^2\}.$$

Here the following denotations have been introduced:

$$f = f^n_{i,j},$$

$$u\Lambda_1 f = (u\Lambda_1 f)_{i,j} = (2h_1)^{-1} [(u + |u|) (f_{i,j} - f_{i-1,j})$$

$$+ (u - |u|) (f_{i+1,j} - f_{i,j})],$$

$$u\Lambda_2 f = (u\Lambda_2 f)_{ij} = (2h_2)^{-1} [(v + |v|) (f_{i,j} - f_{i,j-1})$$

$$+ (v - |v|) (f_{i,j+1} - f_{i,j})],$$

$$\Lambda A = (\Lambda A)_{i,j} = h_1^{-2}(A_{i+1,j} - 2A_{i,j} + A_{i-1,j})$$

$$+ h_2^{-2}(A_{i,j+1} - 2A_{i,j} + A_{i,j-1}).$$

In order to exclude the effects of the 'numerical' magnetic viscosity on reconnection processes of the magnetic lines of force, the terms $(1/2) (h_1 |u| A_{xx} + h_2 |v| A_{yy})$ have been subtracted from the difference equation in the plane case and the terms $(1/2) (h_1 |u| r^{-1} \partial^2 A/\partial r^2 + h_2 |v| A_{zz})$ — in the cylindrical case. Calculations have shown such a procedure to have no effect on stability.

Let us now describe the numerical solution of the plane problem. At the initial moment of time the lines of force of the magnetic field are straight lines parallel to the x-axis. The plasma flow arising with time results in the appearance of the magnetic field component transversal with respect to the plane $y = 0$, and in the formation of the configuration of the 0-type in case of the initial pressure increase in the domain centre, or of the X-type in case of the initial decrease in pressure. Presence of the vertical component of the magnetic field means reconnection of the magnetic lines of force through the neutral surface. Reconnection intensity can be estimated through the value of the magnetic flux passing through the plane $y = 0$, i.e. $\Phi(t) = \int\limits_{0}^{y_{max}} B_y(x,$

$0, t) dx$. An increase in the amplitude of the initial disturbance C results in the growing value of the magnetic flux.

Note, that reconnection of the magnetic lines of force is possible only in the presence of a finite conductivity. Indeed, due to the prohlem symmetry, a plasma flow through the neutral plane is absent and, hence, if the magnetic field is frozen into the plasma ($\sigma = \infty$, $\nu_m = 0$), the magnetic lines of force cannot pass through the neutral surface. A finite-conductivity plasma can be disturbed from the equilibrium state under no external effect whatsoever. It is due to the fact that diffusion results in decreasing gradients of the magnetic field, and the gas-kinetic pressure is not counter-balanced by the magnetic pressure. However, the arising flow remains one-dimensional and no reconnection of the lines of force takes place in the absence of external disturbances.

At the initial moment of time the current $j_z = \mu_0^{-1}(\text{rot } \boldsymbol{B})_z$ is basically concentrated in the vicinity of the neutral plane, then the maximum value of the current density decreases and the current begins occupying a greater part of the domain due to the plasma finite conductivity. The calculation analysis indicates that with growing time there occurs a continuous transfer of the magnetic energy into the thermal and kinetic energies of the system; the reconnection process intensity depending on the conductivity value (or the Reynolds magnetic number Re_m): with reducing Re_m there takes place an increase of the maximum value of the transversal component B_y of the magnetic field in the neutral plane.

Now let us go over to discussing the results of the numerical solution of the cylindrical problem. Under the effect of the magnetic pressure growing on the chamber wall (see the boundary condition (5.17a)), there arises plasma motion. As far as the plasma conductivity is finite, a reconnection of the lines of force takes place through the zero surface of the magnetic field, dynamics of this process depending greatly on the plasma parameters (conductivity, initial density drop) and on the external disturbance (A_1, ω, β).

In the cylindrical case considered the reconnection process can be qualitatively characterized through the difference between the magnetic fluxes via the right ($z = z_{\max}$) and the left ($z = 0$) boundaries of the calculation domain divided into the magnetic flux via the right boundary, i.e.

$$\Phi(t) = (\Phi_1 - \Phi_0)/\Phi_1, \qquad (5.18)$$

where

$$\Phi_0 = 2\pi \int_{r_{00}}^{r_{s0}} rB_z(r, 0, t)\, dr, \quad \Phi_1 = 2\pi \int_{r_{01}}^{r_{s1}} rB_z(r, z_{max}, t)\, dr.$$

Here r_{00} and r_{01} are the coordinates of the potential minimum values on the left and right boundaries, respectively; r_{s0} and r_{s1} are the separatrix radia at $z = 0$ and $z = z_{max}$, respectively. A separatrix is a surface limiting the domain with its total magnetic flux inside equal to zero. Below one can find the $\Phi(t_*)$ values at the moment of time $t_* = 1.5$ depending on the frequency of collective collisions ν_a at the following parameters of plasma and external disturbance: $N = 3$, $\nu_1 = 0$, $A_0 = 3$, $A_1 = 0.2$, $\omega = 0.5$, $\beta_1 = 50$, $r_0 = 0.5$, $\alpha = 15$:

ν_a	10^{-3}	$2,10^{-3}$	$5,10^{-3}$	10^{-2}
$\Phi(t_*)$	0.15	0.35	0.40	0.89.

An increase in the steepness α of the initial distribution of the magnetic field results in decreasing $\Phi(t_*)$ values; with growing radius of the neutral surface r_0 this reconnection characteristic changes in an analogous way.

The considered process of the magnetic field reconnection and diffusion is accompanied by generation of a compression wave and by plasma motion along the axis of a cylindrical chamber. The plasma motion can be traced by way of analysing a time dependence of the plasma mass $M(z, t) = 2\pi \int_0^1 r\varrho(r, z, t)\, dr$ passing through the cross-section of the chamber. The calculation results demonstrate that the wave of compression propagates along the cylinder axis with the velocity approximately equal to $1.2V_A$. Figure 22 presents the dynamics of the magnetic lines of force reconnection and the formation of a closed configuration of the magnetic field.

The finite-difference method discussed above can also be used in other problems of magnetogas-dynamics, for instance, for studying plasma flows in MHD-generators and for calculating various types of plasma instabilities.

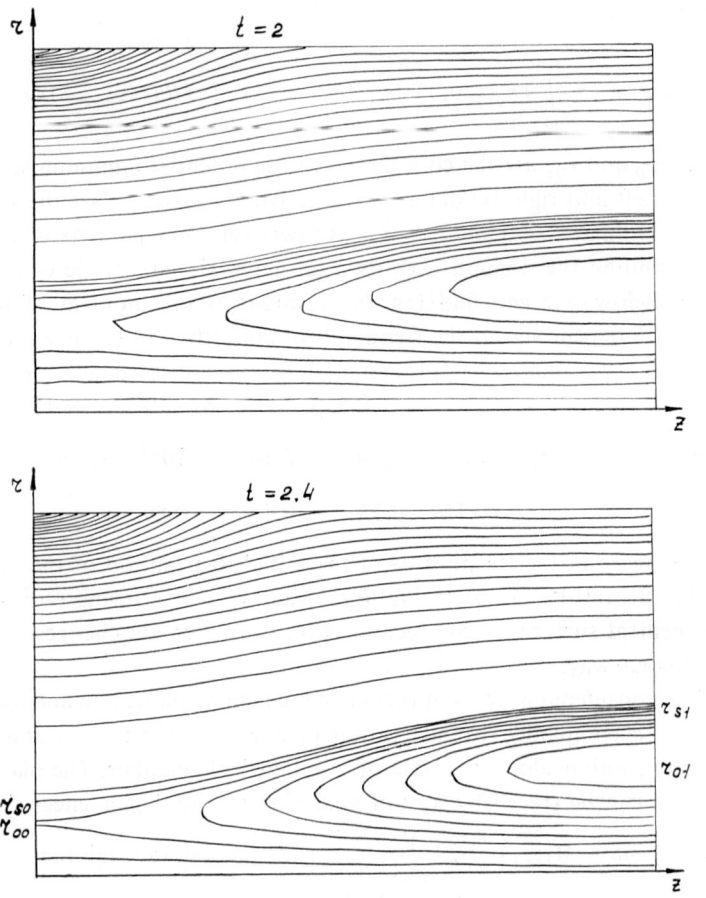

Fig. 22. Configuration of magnetic field lines in the process of reconnection
and a compact torus formation.

Chapter 6

Unsteady Gas-dynamic Processes in Neutron Stars

According to present-day concepts, the character of the final stage of stellar evolution is essentially dependent on the mass of the star entering this phase of its development. Besides the mass, star behaviour in its final phase is also affected by some other factors — rotation, magnetic field, presence of other celestial bodies 'in the vicinity'; their mass, however, being the basic factor. If the initial mass of a star is $M \approx 1 \div 2M_\odot$, and $M_\odot = 2 \times 10^{30}$ kg is the mass of the Sun, then, subsequent to burning out of the nuclear fuel, the inner layers of the star, no longer affected by the counter-pressure, collapse under the effect of gravitation, while the outer layers are ejected into the surrounding space with the velocities of the order of 10^4 km/s — i.e. there occurs a supernona explosion. The greatly collapsed inner layer forms a new equilibrium configuration — a neutron star.

A supposition on the existence of neutron stars was put forward immediately after Chadwick discovered a neutron in 1932, by Landau (1932) and Baade and Zwicki (1934) independently. For a long time a neutron star remained an abstract mathematical construction since the celestial bodies of the sizes of the order of 10 km (by some estimates) and located at distances of dozens of light years, cannot be detected by optical means, even making use of the most powerful telescopes. The situation changed when the radioastronomer Hewish and his co-workers in the Cavendish Laboratory (England) discovered celestial sources of pulse radiation — pulsars (1967). Nowadays it is considered that the most suitable model of a pulsar is a rapidly rotating neutron star with an enormous magnetic field ($\approx 10^6 \div 10^8 T$) and surrounded with a plasma magnetosphere. Thus, neutron stars 'turned into' really existing objects with the mass $1 \div 2M_\odot$, radius R of the order of 10 km and density in the centre $\varrho_c = 2 \cdot 10^{17} \div 4 \cdot 10^{18}$ kg/m^3.

6.1. Equation of State

Theoretical studies of the structure and internal processes in neutron stars are essentially dependent on the relation between the pressure and density of the matter, i.e. on the equation of state. At fixed densities the distances among the neutrons are equal by the order of magnitude to the nucleon sizes; therefore, in order to deduce the equation of state one has to have reliable information on the nuclear forces acting among the nucleons at short distances, which is directly related to the development of experimental and theoretical physics. With the neutron interactions neglected, the simplest equation of state is that of a degenerate neutron gas:

$$p = A_n\{x(2x^2 - 3)(1 + x^2)^{1/2} + 3\ln(x + (1 + x^2)^{1/2})\}$$

$$x = (\varrho/B_n)^{1/3}, \quad A_n = \pi m_n^4 c^5/3h^3, \quad B_n = 8\pi m_n^4 c^3/3h^3. \quad (6.1)$$

In a completely non-relativistic limit, when the density $\varrho \ll 6 \times 10^{18}$ kg/m^3, the neutron gas pressure is

$$p = (1/5)(3\pi^2)^{2/3} \hbar^2 m_n^{-8/3} \varrho^{5/3}.$$

In a completely relativistic limit, which is of a purely mathematical interest, since for it to be achieved the inequality $\varrho \gg 6 \times 10^{18}$ kg/m^3 must hold, from the general formula (6.1) we have

$$p = (1/8)(3\pi)^{1/3} hcm_n^{-4/3} \varrho^{4/3}.$$

A real situation is much more complex since in a superdense matter, account must be taken of the forces of interaction among the neutrons.

In literature a number of equations of state which can be used in studying gas-dynamical processes in neutron stars are being discussed. At low densities ($\varrho \lesssim 10^7$ kg/m³) matter consists of iron nuclei packed in a crystal lattice in such a way that the energy of their Coulomb interaction should be minimum. At $\varrho > 10^7$ kg/m³ the electrons are freed, and at $\varrho \gtrsim 10^{10}$ kg/m³ are completely relativistic. A further increase in density results in the appearance of the nuclei with a greater number of neutrons, and at density $\varrho \approx 4.3 \times 10^{14}$ kg/m³ are enriched with neutrons to such an extent that with growing density there appear free neutrons not trapped in the nuclei (neutron gas). Within the range $4.3 \times 10^{14} \lesssim \varrho \lesssim 2 \times 10^{17}$ kg/m³ the matter consists of nuclei enriched with neutrons, forming a crystal lattice, of a gas of free neutrons and of electrons. If the density $\varrho > 2 \times 10^{17}$ kg/m³, the matter is a mixture of neutrons with an insignificant addition of protons and electrons ($\approx 4\%$). In a neutron star the pressure energy required for calculations is a sum of the energies of isolated nuclei, neutrons and electrons, and the energy of the nuclei bonds in the crystal lattice.

In line with [75], in a neutron matter at densities greater than or of the order of the nuclear density $\varrho_0 \approx 3 \times 10^{17}$ kg/m³ plausible are the π-meson (or pion) condensation and phase transitions with the formation of pion condensates. In this case the total energy density is

$$\varepsilon(\varrho, T) = \varepsilon_N(\varrho) + \varepsilon_\pi(\varrho) + \varepsilon_T(\varrho, T), \qquad (6.2)$$

where ϱ is the matter density. ε_N is the nucleon energy density at zero temperature, ε_π is the energy density of the pion condensate, ε_T is a temperature addition. The pressure p and the matter compressibility K are determined with the formulae $p = \varrho^2 \dfrac{d}{d\varrho}(\varepsilon/\varrho)$, $K = \varrho \dfrac{d^2\varepsilon}{d\varrho^2}$. In [75] the authors suggest the following interpolation formula for the energy pertaining to the formation of a pion condensate:

$$\varepsilon_\pi(\varrho) = -\frac{1}{2}\beta(\varrho)\,(\varrho - \varrho_*)^2\,\delta(\varrho - \varrho_*), \qquad (6.3)$$

where ϱ_* is a certain critical density, $\beta(\varrho) = a + b(\varrho/\varrho_*)^{-1} + c(\varrho_*/\varrho)^{-2}$, a, b, c are the coefficients given together with the ϱ_* values in [75], $\delta(\varrho - \varrho_*) = 1$ at $\varrho > \varrho_*$ and $\delta(\varrho - \varrho_*) = 0$ at $\varrho < \varrho_*$.

The matter compressibility K has a discontinuity at the critical density ϱ_*. Indeed, at $\varrho < \varrho_*$ the compressibility $K(\varrho, T) = K_N + K_T \equiv K_0(\varrho, T)$ is positive. As the density approaches the critical value ϱ_* from the side of high densities, $K \rightarrow K_0 - \beta_0$, where $\beta_0 = a + b + c$. If $\beta_0 > K_0(\varrho_*, T)$, then at densities exceeding the critical one the compressibility is negative, which corresponds to inequality $dp/d\varrho < 0$. The pion condensation results in softening the equation of state, and at certain sets of the parameters ϱ_*, a, b, c this equation turns to non-monotonous (of type of the van der Waals equation of state). The region of a negative compressibility is unstable, therefore, one can expect the appearance of fast processes resulting in an essential reconstruction of the structure of a neutron star due to the arising discontinuity (or jump) of density in the inner layers.

The temperature addition $\varepsilon_T(\varrho, T)$ to the energy density will be expressed in a form corresponding to an idealized degenerate neutron gas. If the energy of thermal excitations T in the system is small as compared to the Fermi energy of the neutron gas $e_F = (\pi^2/2)(3\pi^2)^{2/3} m_n^{-1}(\varrho/m_n)^{2/3}$, where m_n is the neutron mass, then $\varepsilon_T(\varrho, T) = (3\pi^2)^{1/3} (m_n^{2/3}/4\hbar^2) \varrho^{1/3} T^2$. If the inequality $T \ll e_F$ does not hold, the temperature admixture is determined through the general expression:

$$\varepsilon_T(\varrho, T) = (2^{1/2} m_n^{3/2}/\pi^2 \hbar^2) T^{5/2} I_{3/2}(x) -$$

$$- (3/10)(3\pi^2)^{2/3} \hbar^2 m_n^{-2/3} \varrho^{5/3}, \qquad (6.4)$$

where $I_{3/2}(x) = \int_0^\infty z^{3/2}(1 + e^{z-x})^{-1} \, dz, \quad x = e_F/T.$

6.2. Calculations of Neutron Star Models in the Newtonian Approach

Let us consider unsteady gas-dynamical processes in spherically-symmetrical neutron stars with non-monotonous equations of state, which can be derived from expressions (6.2)–(6.4) for energy density. As the governing equations let us choose the equations of the Newtonian non-relativistic theory (NR), which in the mass Lagrangian coordinates are as follows:

$$\frac{\partial u}{\partial t} = -4\pi r^2 \frac{\partial p}{\partial q} - Gq/r^2,$$

$$\frac{\partial r}{\partial t} = u, \qquad \frac{\partial r}{\partial q} = (4\pi r^2 \varrho)^{-1}, \qquad (6.5)$$

$$\frac{\partial}{\partial t}(\varepsilon/\varrho) + p \frac{\partial}{\partial t}(1/\varrho) = -Q,$$

$$\varepsilon = \varepsilon(\varrho, T), \quad p = p(\varrho, T).$$

Here $q(r, t) = \int_0^r 4\pi \varrho(r') r'^2 \, dr'$ is the mass Lagrangian coordinate, u is the macroscopic radial velocity of the matter, G is the gravitational constant, Q are the neutrino energy losses. A relation among the pressure, density and temperature of the matter is determined by the formula $p(\varrho, T) = p_N(\varrho) + p_\pi(\varrho) + p_T(\varrho, T)$, where $p_N(\varrho) = \varrho^2 \frac{d}{d\varrho}[\varepsilon_N/\varrho]$, $p_\pi(\varrho) = \varrho^2 \frac{d}{d\varrho}(\varepsilon_\pi/\varrho)$, $p_T(\varrho, T) = \frac{2}{3}\varepsilon_T(\varrho, T)$.

The characteristic times of temperature decreasing in neutron stars due to neutrino losses by the order of magnitude are equal to $t_\nu \sim 1$ s. Gas-dynamical processes in stars have the time scale $t_H \sim (R^3 GM)^{1/3} \approx 10^5 (\bar\varrho)^{-1/2}$, where $\bar\varrho$ is the mean density of a star. At $\bar\varrho \gtrsim 10^{17}$ kg/m^3 we get $t_H \lesssim 10^{-3}$ s $\ll t_\nu$, therefore, let us study gas-dynamical processes neglecting the cooling of neutron star matter under the effect of neutrino radiation, i.e. setting $Q = 0$. Now let us find an expression relating the temperature and density, for which purpose let us rewrite the equation of energy as $\left(\frac{\partial E}{\partial T}\right)_\varrho \frac{\partial T}{\partial t} + \left[\left(\frac{\partial E}{\partial \varrho}\right)_T - \frac{p}{\varrho^2}\right]\frac{\partial \varrho}{\partial t} = 0,$

$$E = \varepsilon/\varrho,$$

and make use of the thermodynamical equality

$$\left(\frac{\partial E}{\partial \varrho}\right)_T = \varrho^{-2}\left[p - T\left(\frac{\partial p}{\partial T}\right)_\varrho\right].$$

As a result, we get the equation for the temperature

$$\frac{\partial T}{\partial t} = \varrho^{-2} T \frac{(\partial p_T/\partial T)_\varrho}{(\partial E_T/\partial T)_\varrho} \frac{\partial \varrho}{\partial t},$$

wherefrom it can be seen that in the adiabatic case considered, when the neutrino energy losses can be neglected, the temperature distribution is related to the density through the formula

$$T(q, t) = T(q, 0) [\varrho(q, t)/\varrho(q, 0)]^{2/3}. \tag{6.6}$$

To complete the formulation of the problem, equations (6.5), (6.6) must be supplemented with the initial and boundary conditions.

A star can enter the domain of non-monotonous pressure dependence on density in a number of ways, for instance, at increasing mass and density due to matter accretion, at collapsing in the process of formation, or at cooling. Let us set the initial density distribution as

$$\varrho(r, 0) = \varrho_c^0[1 - [r/R_0)^s], \quad s = 3[(M_0/M) - 1]^{-1}, \tag{6.7}$$

$$M = (4\pi/3) \varrho_c^0 R_0^3,$$

where ϱ_c^0 is the initial density in the star centre, R_0 is the initial radius, M is the star mass. The values of the Euler coordinate $r^0(q)$ at the moment of time $t = 0$ are determined by the iteration method from the equation $dr^0/dq = [4\pi\varrho(r^0, 0) (r^0)^2]^{-1}$. The boundary conditions at $q = 0$ and $q_{max} = M$ are given in a standard way:

$$u(0, t) = r(0, t) = p(M, t) = 0. \tag{6.8}$$

Problems (6.5)–(6.8) have been numerically solved [13, 25, 14] using the explicit finite-difference scheme on staggered meshes, which is analogous to the scheme discussed in Chapter 3, with a natural generalization for the spherical case. As far as there is no dissipation and physical considerations prompt that there may arise a density discontinuity, the difference scheme is modified with an artificial quadratic Neumann–Richtmyer viscosity.

$$u_j^{n+1/2} = u_j^{n-1/2} - (4\pi\tau/h) (r_j^n)^2 (\bar{p}_{j+1/2}^n - \bar{p}_{j-1/2}^n) - \tau G q_j (r_j^n)^{-2},$$

$$r_j^{n+1} = r_j^n + \tau u_j^{n+1/2}, \quad \varrho_{j-1/2}^{n+1} = (3h/4\pi) [(r_j^{n+1})^3 - (r_{j-1}^{n+1})^3],$$

$$\bar{p}_{j-1/2}^{n+1} = p_{j-1/2}^{n+1} - v\varrho_{j-1/2}^{n+1} |u_j^{n+1/2} - u_{j-1}^{n+1/2}| (u_j^{n+1/2} - u_{j-1}^{n+1/2}),$$

$$T_{j-1/2}^{n+1} = T_{j-1/2}^0 (\varrho_{j-1/2}^{n+1}/\varrho_{j-1/2}^0)^{2/3},$$

$$p_{j-1/2}^{n+1} = p(\varrho_{j-1/2}^{n+1}, T_{j-1/2}^{n+1}), \quad j = 1, \dots, J - 1.$$

In the difference equations (6.9) τ is a time step, h is a step along the mass Lagrangian coordinate, v is the coefficient of artificial viscosity. At the boundary point $j = J$ the difference equation of motion is written as

$$u_J^{n+1/2} = u_J^{n-1/2} + (4\pi\tau/h) (r_J^n)^2 (3\bar{p}_{J-\frac{1}{2}}^n - \tfrac{1}{3} \bar{p}_{J-3/2}^n) -$$
$$- \tau G q_J (r_J^n)^{-2}.$$

Let us now describe the results of a series of calculations [13, 25], obtained by the above scheme at the following parameters: $\varrho_c^0 = 3 \times 10^{17}$ kg/m³, $M = 0.45, 1.0, 1.4, 1.85, 2.0 M_\odot$, $R_0 = 13 \div 18$ km, $T_c^0 = 0, 50, 150 \times 10^9$ K (the initial temperature in the star centre). For the monotonous part $\varepsilon_N(\varrho)$ of the equation of state, use was made of the equation of state from [4] in the density region from 10^7 to 7.7×10^6 kg/m³ and from [84] in the density region from 8.4×10^{16} to 1.26×10^{19} kg/m³ for a purely neutron matter. The non-monotonous part was set by the parameters $a = 0.91$, $b = -0.23$, $c = 0.14$ (in pion units); the critical density $\varrho_* = 4.1 \times 10^{17}$ kg/m³.

At the moment $t = 0$ the whole of a star (or stars, as calculations were carried out for various masses) is completely included in the domain of monotonous pressure dependence on density, since $\varrho_c^0 < \varrho_*$. The initial configurations are unstable and, therefore, the mass layers acquire the velocities directed to the centre, and with time the star 'enters' the domain of non-monotonous dependence of pressure on density. In a 'cold' matter, which corresponds to the temperature $T_c < 10^{10}$ K (in fact, the temperature in such cases can be set equal to zero, as temperature additions are negligibly small), in all the calculation variants at the times $\lesssim 1$ ms counted from the moment of the beginning of a collapse, there arises and gradually stabilizes a density discontinuity against a continuous pressure. It results in a re-arrangement of the star structure: there is formed a dense core and a less dense shell, divided by a sharp boundary. Figure 23 presents the density distributions in a neutron star with the mass $M = 1.4M_\odot$ depending on the Lagrange coordinate at the moments $t = 0$ and $t = 3.7$ ms, dashed line denoting the density discontinuity. This density drop occurs at one step h of the Lagrangian grid irrespective of the step length, since the difference scheme (6.9) reproduces contact discontinuities without any distortion (see, for instance, [92]). The radii

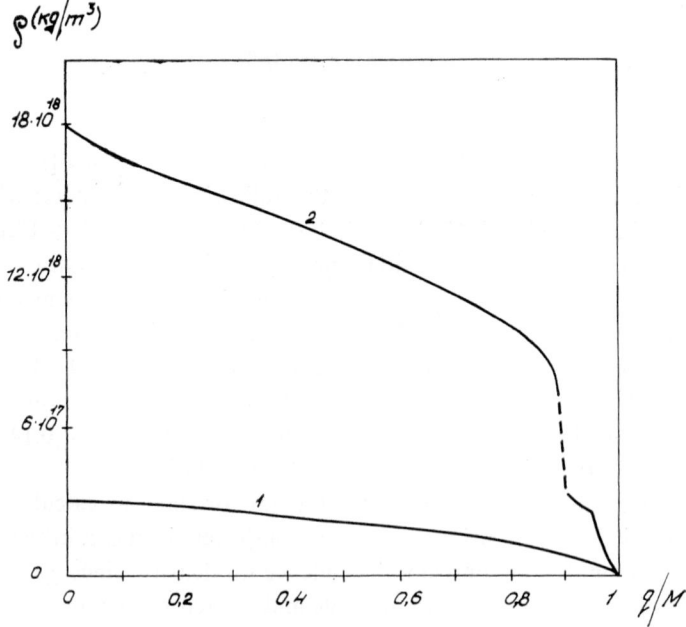

Fig. 23. Density distribution in a star. $1 - t = 0.2 - t = 3.7$ ms.

of the stars R and dense cores R_* with time get subjected to insignificant fluctuations from some mean values, which is reflected in the table below:

$M/M_\odot$	0.45	1.0	1.4	1.85	2.0
R_{km}	8.4	8.65	9.1	9.5	9.55
R_{*km}	5.15	6.45	7.7	8.3	8.55

Calculations carried out at some other initial configurations and employing other ways of 'introducing' the star into the region of critical densities (and of a non-monotonous equation of state) demonstrate that in the process of quasi-establishment the radii R, R_* and the densities in the star centre and in the discontinuity are defined solely by the star mass M and are independent of the initial conditions. The above calculations were carried out on the grid with the step $h = M/60$, the total energy preserved in all the variants by not greater than 0.5%.

Since in forming a pion condensate there occurs a sharp division of the star with respect to density into a core and a shell, it is quite plausible that in this case the shell of a neutron star can be ejected away. To verify this hypothesis, put forward in [76], the author of the present monograph has carried out calculations for neutron stars with $M = 0.45$, 1.4, $1.85 M_\odot$ on the grid with reducing from the centre to periphery steps $h = 5 \times 10^{-2} \div 10^{-5} M$. Figure 24 shows the time dependence of the radii corresponding to different values of the mass Lagrange coordinate for the variant $M = 0.45 M_\odot$, $R_0 = 26$ km. In this case the shell mass is $0.32M$, and approximately $0.015M$ is ejected away, the energy of the ejected part being less than 10^{43} J. Calculations indicate that for stars with the mass $M = 1.4$ and $1.85 M_\odot$ the ejection is plausible, but this value is far less than $10^{-5}M$ (the masses of the star shells are equal to $0.083M$ and $0.067M$, respectively). Analysis of these numerical results leads one to the conclusion that in the considered models of neutron stars the ejection of an extremely small part of the mass is not conditioned by the formation of a dense core and density discontinuity, but is due to non-equilibrium of the initial configurations.

In a hot matter with the temperature $T \gtrsim 5 \times 10^{10}$ K, as was the case for a cold matter, there is formed a density discontinuity provided the conditions of pion condensation are met. The table below presents the mean in time values of $\bar{R}$, $\bar{R}_*$, of the core mass M_*, of the density and temperature in the star centres, as well as the density values on the discontinuity ϱ_1, ϱ_2 for various initial temperatures T_c^0.

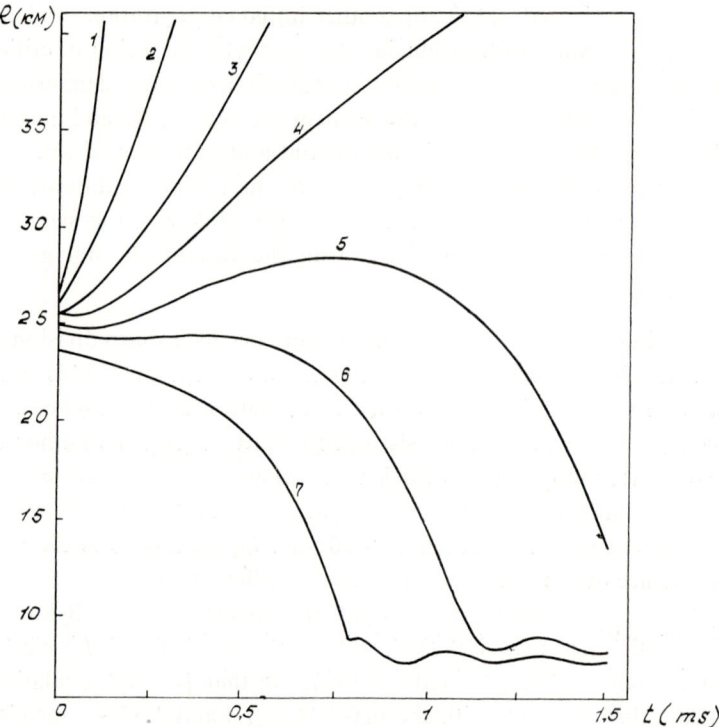

Fig. 24. Time dependence of the Euler coordinates for a star with $M =$ 0.45 $M_\odot$. The curves correspond to the following values of the mass Lagrange coordinates: $1 - M$, $2 - 0.998\ M$, $3 - 0.988\ M$, $4 - 0.978\ M$, $5 - 0.968\ M$, $6 - 0.958\ M$, $7 - 0.940\ M$.

The amplitude and period of fluctuations in the star radius and the core radius increase with temperature, Thus, in the variant with $M = M_\odot$ the relative amplitude of the star radius fluctuations is 2% at $T_c^0 = 5 \times 10^{10}$ K and 15% at $T_c^0 = 15 \times 10^{10}$ K. With growing mass the relative amplitude of the radius R fluctuations reduces. With growing initial temperature T_c^0 there occurs an increase in the star radius, a decrease in the density in the centre, in the mass M_* and in

$\dfrac{M}{M_\odot}$	$T_c^0(10^{10}$ K)	$\bar{R}_{km}$	$\bar{R}_{*km}$	$\dfrac{M_*}{M_\odot}$	$\varrho_c(10^{18}$ kg/m³)	$T_c(10^{10}$ K)	$\varrho_1(10^{17})$	$\varrho_2(10^{17})$
	0	8.6	7	0.9	1.65	0	3.5	7.9
1	5	8.8	7	0.9	1.63	15.5	3.7	7.9
	10	10	7.3	0.8	1.52	44.6	4.4	6.7
	0	9.1	7.8	1.3	1.81	0	3.1	7.3
1.4	5	9.1	7.8	1.3	1.78	16.4	3.1	7.3
	10	10	7.8	1.2	1.68	47.3	3.6	7.2

the radius R_* of the core. With growing initial temperature the value of a density jump decreases. As was the case for a cold matter, calculations with the finite initial temperature demonstrate that no shell ejection takes place when a dense core is formed.

6.3. Dynamics of Neutron Stars Taking into Account the General Relativity Effects

The calculations carried out within the Newtonian non-relativistic approach (NR) have shown that in the equilibrium state a star with the mass equal, for instance, to the mass of the Sun, has at the chosen above equation of state the radius $R = 8.6$ km, which is not too great as compared to the Schwarzschild gravitational radius for such a mass $R_g \approx 3$ km. Let us therefore consider gas-dynamical processes in neutron stars with the mass $M \gtrsim M_\odot$ on the basis of the General Relativity Theory (GR).

According to [12], let us choose the Einstein equations as governing in the spherically-symmetrical co-moving system of coordinates with the metrics

$$ds^2 = a^2(q, t) c^2 dt^2 - b^2(q, t) dq^2 - r^2(q, t) d\Omega^2 \quad (6.10)$$

where $d\Omega^2 = d\theta^2 + \sin^2 \theta \, d\varphi^2$; the coordinate r is determined in such a way that the circumference centred at the origin of the coordinates equals $2\pi r$; the coordinate q, which by analogy with general hydrodynamics can be termed as Lagrangian, is defined as a rest mass of the matter in a sphere of the radius q, i.e. $q = \int\limits_0^q 4\pi\varrho r^2 \dfrac{\partial r}{\partial q'} \, dq'$,

ϱ is the density of the rest mass; at such a choice of the coordinate the metrical coefficient $b = (4\pi\varrho r^2)^{-1}$. In this case the Einstein equations can be presented in the form resembling the equations of gasdynamics, i.e.

$$\frac{\partial u}{\partial t} = -a \left(4\pi r^2 \Gamma \frac{\partial p}{\partial q} w^{-1} + Gmr^{-2} + 4\pi Gc^{-2}pr \right),$$

$$\frac{\partial r}{\partial t} = au, \quad (\varrho r^2)^{-1} \frac{\partial}{\partial t} (\varrho r^2) = a \frac{\partial u}{\partial q} \left(\frac{\partial r}{\partial q} \right)^{-1},$$

$$\frac{\partial E}{\partial t} + p \frac{\partial}{\partial t} (1/\varrho) = 0, \qquad\qquad (6.11)$$

$$(aw)^{-1} \frac{\partial}{\partial q} (aw) = (wc^2)^{-1} \left(\frac{\partial E}{\partial q} + p \frac{\partial}{\partial q} (1/\varrho) \right),$$

$$p = p(\varrho),$$

where u is the radial velocity, E is the internal energy per kg of matter, $w = 1 + (E + p/\varrho) c^{-2}$, $m = 4\pi \int\limits_0^q \varrho(1 + Ec^{-2}) r^2(\partial r/\partial q') \, dq'$, is the total mass of the matter in a sphere of the 'radius' q, $\Gamma = 4\pi\varrho r^2 (\partial r/\partial q)^2$.

The simplest way of setting an initial configuration is to choose a star with a uniform density distribution, which has been the case, since the calculations discussed in the preceding paragraph demonstrate practical independence of a star evolution of the initial conditions. The boundary conditions are set in the following way:

$$r = 0, \quad u = 0, \quad \Gamma = 1 \quad \text{at} \quad q = 0, \qquad (6.12)$$

$$p = 0, \quad a = \bar{a} \quad \text{at} \quad q = M.$$

Setting a boundary condition for the coefficient a at $q = M$ is to be done with caution; a way of its setting will be given below when describing the algorithm of the numerical solution of equations (6.11).

As was the case in the preceding paragraph, let us make use of the scheme on the shifted grids with the artificial quadratic Neumann–Richtmyer viscosity:

$$u_t = -a(4\pi r^2 \Gamma w^{-1}\bar{p}_q + Gmr^{-2} + 4\pi Gc^{-2}r\bar{p}),$$

$$r_t = a\hat{u},$$

$$\hat{\varrho} = \varrho r^2 \hat{r}^{-2} \exp(-\tau a\hat{u}_{\bar{q}}\hat{r}_{\bar{q}}),$$

$$E_t = -\hat{\bar{p}}(1/\varrho)_t,$$

$$\hat{\bar{p}} = p(\hat{\varrho}) + \hat{Q}, \qquad\qquad (6.13)$$

$$\hat{w} = 1 + (\hat{E} + \hat{\bar{p}}/\hat{\varrho})\,c^{-2},$$

$$\hat{a} = aw\hat{w}^{-1}\exp(\hat{\varphi}),$$

$$\hat{\varphi} = \hat{w}^{-1}\{c^{-2}(\hat{E}_{\bar{q}} + \hat{\bar{p}}(1/\hat{\varrho})_{\bar{q}}),$$

$$\hat{Q} = \nu\hat{\varrho}\,[(\hat{r}^2\hat{u})_{\bar{q}}]^2\,\Gamma^{-1}r^{-4}.$$

Now let us consider the problem of setting a boundary condition for the metric coefficient a at $q = M$. To determine this coefficient in the last but one point $j = J - \frac{1}{2}$ at any moment of time, let us integrate the fifth equation of system (6.11) in the limits from $(aw)_{J-\frac{1}{2}}$ to $(aw)_{J+\frac{1}{2}}$ and get

$$a_{J-\frac{1}{2}} = (aw)_{J+\frac{1}{2}}\,(w_{J-\frac{1}{2}})^{-1}\exp\{-c^{-2}w_J^{-1}[E_{J+\frac{1}{2}} - E_{J-\frac{1}{2}}$$
$$+ p_J(1/\varrho_{J+\frac{1}{2}} - 1/\varrho_{J-1/2})]\},$$

wherefrom we have to the accuracy of the values of the second order of smallness

$$a_{J-\frac{1}{2}} = (w_{J-\frac{1}{2}})^{-1}\,a_{J+\frac{1}{2}}.$$

Setting $a_{J+\frac{1}{2}} = 1$ will denote synchronization of the coordinate and real time at the boundary point. Such an assumption is valid if the star radius is sufficiently great as compared to the Schwarzschild radius. If it is not the case, then, since the point $j = J + \frac{1}{2}$ is beyond the stellar matter, there should be introduced in it the Schwarzschild metrics for a gravitational field in vacuum induced by the centre-symmetrical mass distribution, i.e.

$$ds^2 = (1 - R_g/r)\, c^2\, dt^2 - (1 - R_g/r)^{-1}\, dr^2 - r^2\, d\Omega^2 .$$

Comparing with metrics (6.10) indicates that (6.10) transforms continuously into the Schwarzschild metrics, provided we set $\bar{a} = 1 - R_g/R$. Therefore, the boundary condition for the metric coefficient a is as follows:

$$a_{J-\frac{1}{2}} = (1 - 2GM/Rc^2)\, (w_{J-\frac{1}{2}})^{-1} .$$

The rest boundary and initial conditions are trivially written in a finite-difference form.

For a qualitative verification of the algorithm there have been carried out calculations of the problem on a collapse of the dust-like sphere which, as is known, has an exact analytical solution at $\bar{p} = 0$, $a = 1$:

$$t(8\pi G\varrho_0/3)^{1/2} = f^{1/2}(1 - f)^{1/2} + \arcsin (1 - f)^{1/2},$$

where $f = r(q, t)/r(q, 0)$. As shown by calculations, the value of the relative error in determining the coordinate r did not exceed 10^{-6}, and in determining the time of the total collapse $t_c = \frac{1}{2}\pi R^{3/2}(2GM)^{-1/2}$ it did not exceed 10^{-5}.

Let us describe the results of a series of calculations [12] carried out by the finite-difference scheme (6.13). The initial configuration is a uniform star with the mass $M = M_\odot$, density $\varrho(q, 0) = 9.6\times 10^{15}$ kg/m³, radius $R_0 = 36.7$ km, internal energy $E(q, 0) = 1.56\times 10^{11}$ J/kg, pressure $p(q, 0) = 1.17\times 10^{30}$ N/m². The initial velocities of the star matter were set proportional to the local parabolic velocity $u(q, 0) = 0.7\,(GM/r(q, 0))^{1/2}$. The equation of state was also chosen non-monotonous, as in the NR model. The grid was taken with the steps $\Delta r = R_0/60$, $\Delta q_j \equiv h_j = 10^{-5} \div 10^{-2}M$.

Calculations of the dynamics of a neutron star with the mass equal to that of the Sun carried out in the NR and GR approaches at the fixed initial and boundary conditions (the same in both models) have shown that in a qualitative respect both models give similar results.

Setting a uniform initial configuration does not put a limit to their conformity in numerical investigation. In the NR model the analogous problem was being solved at various initial configurations, including the equilibrium ones, which correspond to the monotonous equation of state.

Numerical solutions demonstrate that differences in star dynamics are observed only within a comparatively short time from the moment of the origination of a collapse, and the disintegration of the initial discontinuity does not distort the parameters of the shock wave arising at the collapse stage moving outward from the centre. To both sides of this wave front the matter is in different phase stages. When approaching the peripheral zones, it disintegrates into a shock wave to both sides of the front of which the matter is in the same phase stage (normal, but not superdense), and into a gradually stabilizing contact discontinuity.

Both in the NR and GR cases the quasi-equilibrium (with small fluctuations) configuration is a dense core and a less dense shell; the density values on the discontinuity correspond to the non-monotonous domain of the equation of state. Under quasi-steady conditions the density in the centre ϱ_c and the core mass M_* appear greater when the GR effects are allowed for, while the core and shell sizes are less than in the NR model. The corresponding values are given below.

	NR	GR
ϱ_c, 10^{18} kg/m^3	1.7	2.1
R_{km}	8.6	8.2
R_{*km}	6.9	6.4
$M_*/M_\odot$	0.87	0.72

In the above calculations the total mass proves to be less than the rest mass: it equals $0.983 M_\odot$ at $t = 0$ and $0.458 M_\odot$ when the quasi-equilibrium state is reached. No shell ejection, which could be considered a result of the phase transition followed by the formation of a pion condensate, takes place. It is to be noted that the method presented above of the numerical solution of GR equations in the co-moving system of coordinates makes it possible to study the dynamics of spherically-symmetrical stars of various masses, including the investigation of gravitational collapse.

By way of conclusion it should be noted that numerical simulation is becoming a more and more widely spread method for studying complex scientific and technological problems. One should, however, always remember about thorough selection of the models, and algorithms and carry out an ultimate comparison with the available analytical solutions.

References

1. Abe K. and Inoue O. Fourier expansion solution of the Korteweg–de Vries equation. *J. Comp. Phys.* 1980, **34** (2), 202–210.
2. Alikhanov S. G., Alinovsky N. I., Dolgov-Saveliev G. G. *et al.* Development of the programme on collisionless shock waves. In: *Plasma Physics and Contr. Nucl. Fusion Res.* IAEA, *Vienna*, 1969.
3. Al'pert Ya. L. *Waves and Satellites in Space Plasma* (in Russian). Nauka, Moscow, 1974.
4. Baym G., Pethick C. and Sutherland P. The ground state of matter at high densities: equation of state and stellar models. *Astrophys. J.* 1971, **170** (2), 239–317.
5. Bellman R. and Kalaba R. *Quasilinearization and Non-Linear Boundary Value Problems.* Elsevier Publishing Company, Inc., New York, 1965.
6. Berezin Yu. A. Non-linear motions in anisotropic plasma. *Soviet Physics JETP*, 1972, **34** (5), 998–1000.
7. Berezin Yu. A. *Numerical Investigation of Non-linear Waves in a Rarefied Plasma* (in Russian). Nauka, Novosibirsk, 1977.
8. Berezin Yu. A. On formation of solitons (in Russian). *Zh. Tekh. Fiz* 1968, **38** (1), 24–27.
9. Berezin Yu. A. On non-linear stage of firehose instability (in Russian). *Prikl. Mekh. i Tekh. Fiz.* 1970, (3), 3–10.
10. Berezin Yu. A. On numerical solutions of the Korteweg–de Vries equation (in Russian). *Chisl. Met. Mekh. Spl. Sr., Novosibirsk,* 1973, **4** (2), 20–31.

11. Berezin Yu. A. On waves of finite amplitude in a rarefied plasma (in Russian). *Prikl. Mekh. i Tekh. Fiz.* 1965, (5), 116–118.

12. Berezin Yu. A. and Dmitrieva O. E. Dynamics of a developing neutron star in general relativity. *Sov. Astron. Lett.* 1984, **10** (3), 175–176.

13. Berezin Yu. A., Dmitrieva O. E. and Yanenko N. N. Calculations of model neutron stars with a pion condensation. *Sov. Astron. Lett.* 1982, **8** (1), 43–45.

14. Berezin Yu. A., Dmitrieva O. E. and Yanenko N. N. Modelling gas-dynamic processes in neutron stars with a phase transition. *Lect. Notes in Phys.* 1982, (170), 138–142.

15. Berezin Yu. A. and Dudnikova G. I. Dynamics of neutral layers in plasma (in Russian). *Prikl. Mekh. i Tekh. Fiz.* 1982, (3), 14–17.

16. Berezin Yu. A. and Dudnikova G. I. Unsteady waves propagating along a magnetic field in plasma (in Russian). *Prikl. Mekh. i Tekh. Fiz.* 1971, (1), 141–144.

17. Berezin Yu. A., Dudnikova G. I., Eselevitch V. G. and Kurtmullaev R. Kh. Structure of an oblique shock wave at high Mach numbers (in Russian). *Prikl. Mekh. i Tekh. Fiz.* 1969, (4), 100–104.

18. Berezin Yu. A. and Karpman V. I. Nonlinear evolution of disturbances in plasma and other dispersive media. *Soviet Phys. JEPT*, 1967, **24**, 1049–1056.

19. Berezin Yu. A. and Karpman V. I. On the theory of unsteady surface waves (in Russian). *Prikl. Mekh. i Tekh. Fiz.* 1964, (5), 135–137.

20. Berezin Yu. A. and Karpman V. I. On the theory of unsteady waves of finite amplitude in rarefied plasma (in Russian). *Zh. Exper. i Teor. Fiz.*, 1964, **46** (5), 1880–1890.

21. Berezin Yu. A. and Khenkin P. V. Expansion of a plasma cylinder through plasma background (in Russian). *Prikl. Mekh. i Tekh. Fiz.* 1982, (6), 37–40.

22. Berezin Yu. A., Kovenya V. M. and Yanenko N. N. Implicit

numerical method for the blunt-body problem in supersonic flows. *Computers and Fluids* 1975, **2** (2/3), 271–283.

23. Berezin Yu. A., Kovenya V. M. and Yanenko N. N. Numerical solutions of the problems of magnetic gas dynamic flows round the bodies. *Lect. Notes in Phys.* 1975, (35), 85–90.

24. Berezin Yu. A., Kurtmullaev R. Kh. and Nesterikhin Yu. E. Collisionless shock waves in a rarefied plasma (in Russian). *Fiz. Gor. i Vzr.* 1966, **1** (1), 3–28.

25. Berezin Yu. A., Mukanova B. G. and Fedoruk M. P. The effect of non-zero temperature upon the dynamics of neutron stars with pion condensation. *Sov. Astron. Lett.* 1983, **9** (1), 63–65.

26. Berezin Yu. A. and Sagdeev R. Z. On the theory of non-linear waves in plasma (in Russian). *Prikl. Mekh. i Tekh. Fiz.* 1966, (2), 3–6.

27. Berezin Yu. A. and Sagdeev R. Z. One-dimensional non-linear model of anisotropic plasma instability (in Russian). *Dokl. AN SSSR*, 1969, **184** (3), 570–573.

28. Berezin Yu. A. and Vshivkov V. A. *Method of Particles in the Dynamics of a Rarefied Plasma* (in Russian). Nauka, Novosibirsk, 1980.

29. Berezin Yu. A. and Vshivkov V. A. On critical parameters of shock waves in plasma (in Russian). *Prikl. Mekh. i Tekh. Fiz.* 1976, (2), 27–36.

30. Berezin Yu. A. and Vshivkov V. A. On the firehose instability of Alfven waves. *J. Comp. Phys.* 1976, **20** (1), 81–96.

31. Berezin Yu. A. and Vshivkov V. A. Shock waves with an arbitrary amplitude in a rarefied plasma with a magnetic field. (in Russian) *Fizika Plasmy* 1977, **3** (2), 365–370.

32. Berezin Yu. A., Vshivkov V. A. and Dudnikova G. I. Structure of unsteady switch-on waves (in Russian). *Prikl. Mekh. i Tekh. Fiz.* 1976, (5), 58–60.

33. Berezin Yu. A., Vshivkov V. A. and Krygin V. D. On two-dimensional wave processes in a dispersive medium. (in Russian) *Prikl. Mekh. i Tekh. Fiz.* 1980, (6), 50–53.

34. Birdsall C. K. and Fuss D. Clouds-in-clouds, clouds-in-cell

physics for many-body plasma simulation. *J. Comp. Phys.* 1969, **3** (4), 494–511.

35. Birdsall C. K., Langdon A. B. and Okuda H. Finite-size par- ticle physics applied to plasma simulation. In: *Methods in Com- putational Physics*, eds. B. Alder, S. Fernbach and M. Roten- berg. Academic Press, New York, 1970, vol. 9, 241–258.

36. Braginskii S. I. Transport phenomena in plasmas. In: *Reviews of Plasma Physics*. Consultants' Bureau, New York, 1965, vol. 1, p. 205.

37. Brushlinsky K. V., Zaborov A. M. and Syrovatsky S. I. Nu- merical analysis of a current layer in the vicinity of a magnetic zero line (in Russian). *Fizika Plazmy* 1980, **6** (2), 297–311.

38. Canosa J. and Gasdag J. The Korteweg–de Vries–Burgers equation. *J. Comp. Phys.* 1977, **23** (4), 393–403.

39. Chen L. and Okuda H. Theory of plasma simulation using multi-pole-expansion scheme. *J. Comp. Phys.* **19** (4), 339 –352.

40. Chew G. F., Goldberger M. L. and Low F. E. Boltzmann equation and hydromagnetic collisionless equations for a single fluid. *Proc. R. Soc.* 1956, **A 236** (1), 112–131.

41. Chodura R. A hybrid fluid-particle model of ion-heating in high-Mach-number shock waves. *Nucl. Fusion* 1975, **15** (1), 55–63.

42. Dyatchenko V. F. and Imshennik V. S. A converging cylin- drical wave in the plasma with the front structure accounted for (in Russian). *Zh. Vytch. Mat. i Matem. Fiz.* 1963, 3 (5), 915–926.

43. Dyatchenko V. F. and Imshennik V. S. Computation results of the MHD model of a non-cylindrical z-pinch (in Russian). IPM AN SSSR, Moscow, 1973, Preprint N 40.

44. Dyatchenko V. F. and Imshennik V. S. On the MHD theory of a pinch-effect in a high-temperature dense plasma (in Rus- sian). In: *Problems of Plasma Theory*. Atomizdat, Moscow, 1967, (5), 394–438.

45. Eselevitch V. G. Experimental studies of collisionless oblique

shock waves (in Russian). Cand. Thesis, IYaF SO AN SSSR, Novosibirsk, 1970.

46. Eselevitch V. G., Es'kov A. G., Kurtmullaev R. Kh. and Malyutin A. I. Fine structure of shock waves in plasma and the mechanism of ion-acoustic turbulence saturation (in Russian). *Zh. Eksp. i Teor. Fiz.* 1971, **60** (5), 1658–1671.

47. Eselevitch V. G., Es'kov A. G., Kurtmullaev R. Kh. and Malyutin A. I. Isomagnetic jump in a collisionless shock wave (in Russian) *Zh. Eksp. i Teoret. Fiz.* 1971, **60** (6), 2079–2091.

48. Es'kov A. G., Kurtmullaev R. Kh., Malyutin A. I. *et al.* Liner compression of a toroidal high β-plasma. *Proc. of III Topical Conf. on High β-Plasmas*, 1975, Culham, England.

49. Es'kov A. G., Kurtmullaev R. Kh., Malyutin A. I. and Pil'sky V. I. Studies on the character of turbulent processes in a shock waves front in plasma (in Russian). *Zh. Eksp. i Teor. Fiz.* 1969, **56** (5), 1480–1491.

50. Galeev A. A. and Sagdeev R. Z. Non-linear plasma theory (in Russian). In: *Problems of Plasma Theory.* Atomizdat, Moscow, 1973, (7), 3–145.

51. Gardner G. S., Greene J. M., Kruskal M. D. and Miura R. M. Method for solving the Korteweg–de Vries equation. *Phys. Rev. Lett.* 1967, **19** (19), 1095–1097.

52. Gasenko V. G. Chart of the BKV equation solutions (in Russian). In: *Studies on Hydrodynamics and Heat Exchange.* IT SO AN SSSR, Novosibirsk, 1976.

53. Gjevik B. Occurence of finite-amplitude surface waves on falling liquid films. *Phys. Fluids* 1970, **13** (8), 1918–1925.

54. Hain K., Hain G., Roberts K. V. *et al.* Fully ionized pinch collapse. *Z. Naturforsch.* 1960, **15a** (12), 1039–1051.

55. Harris E. G. On a plasma sheath separating regions of oppositely directed magnetic fields. *Nuovo Cim.* 1962, **23** (1), 115–121.

56. Hockney R. W. A computer experiment of anomalous diffusion. *Phys. Fluids* 1966, **9** (9), 1826–1835.

57. Imshennik V. S. Two-dimensional unsteady numerical plasma

models of the problem of stretching in z-pinches and plasma
focus (in Russian). In: *Numerical Methods of Plasma Physics*.
Nauka, Moscow, 1977, 100–121.

58. Iskol'dsky A. M., Kurtmullaev R. Kh., Nesterikhin Yu. E. and
Ponomarenko A. G. Experiments on collisionless shock waves
in plasma (in Russian). *Zh. Eksp. i Teor. Fiz.* 1964, **47** (2) (8),
774–776.

59. Jeffrey A. and Kakutani T. Weak non-linear dispersive waves.
SIAM Rev. 1972, **14** (14), 582–643.

60. Kadomtsev B. B. *Collective Phenomena in Plasma* (in Russian).
Nauka, Moscow. 1976.

61. Kadomtsev B. B. and Petviashvili V. I. On stability of solitons
in weak dispersive media (in Russian). *Dokl. AN SSSR* 1970,
192 (4), 753–756.

62. Karpman V. I. *Non-linear Waves in Dispersive Media*. Pergamon
Press, New York, 1975.

63. Kennel C. E. and Sagdeev R. Z. Collisionless shock wave in
high β-plasmas. *J. Geophys. Res.* 1967, **72** (13), 3303–3341.

64. Khalatnikov I. M. *Theory of Superfluidity* (in Russian). Nauka,
Moscow, 1971.

65. Korteweg D. J. and de Vries G. On the change of form of long
waves advancing in a rectangular channel and on a new type of
long stationary waves. *Phil. Mag.* 1895, **39** (5), 422–443.

66. Kruer W. L., Dawson J. M. and Rosen B. The dipole expansion method for plasma simulation. *J. Comp. Phys.* 1973, **13**
(1), 114–129.

67. Kurtmullaev R. Kh., Masalov K. I. and Semenov V. N. Shock
waves propagating along a magnetic field in a collisionless
plasma (in Russian). *Zh. Eksp. i Teor. Fiz.* 1971, **60** (1), 400–407.

68. Kurtmullaev R. Kh., Nesterikhim Yu-E., Pil'sky V. I. and Sagdeev R. Z. Mechanism of plasma heating by collisionless shock
waves. In: *Plasma Phys. and Contr. Nucl. Fusion Res.* IAEA,
Vienna, 1966, 47–68.

69. Kuznetsov V. V., Nakoryakov V. E., Pokusaev B. G. and Shreiber I. R. A fluid with gas bubbles as an example of the Korte-

weg–de Vries–Burgers medium (in Russian). *Pis'ma v ZETF* 1976, **23** (4), 194–198.

70. Kuznetsov V. V., Nakoryakov V. E., Pokusaev B. G. and Shreiber I. R. Propagation of perturbations in a gas–liquid mixture. *J. Fluid Mech.* 1978, **85** (1), 85–96.

71. Landau L. D. and Lifshitz E. M. *Fluid Mechanics* (Course of Theoretical Physics, Vol. 6). Pergamon Press, Oxford, 1959.

72. Lewis H. R., Sykes A. and Wesson J. A. A comparison of some particle-in-cell plasma simulation methods. *J. Comp. Phys.* 1972, **10** (1), 85–106.

73. Liewer P. C. and Krall N. A. Anomalous penetration of a magnetic pulse into a plasma. *Phys. Rev. Lett.* 1973, **30** (25), 1242–1245.

74. Maurin L. N. and Totchigin A. A. Solitons on a falling liquid film (in Russian). *Prikl. Mech. i Tech. Fiz.* 1979, (4), 47–54.

75. Migdal A. B. Pion fields in nuclear matter. *Rev. Mod. Phys.* 1978, **50** (1), 107–172.

76. Migdal A. B., Chernoutsan A. I. and Mishustin I. N. Pion condensation and dynamics of neutron stars. *Phys. Lett.* 1979, **83b** (2), 158–160.

77. Moodie T. B. and Haddow J. B. Dispersive effects in wave propagation in thin-walled elastic tubes. *J. Acoust. Soc. Am.* 1978, **64** (2), 522–528.

78. Morse R. L. and Nielson S. W. Numerical simulation of warm two-beam plasma. *Phys. Fluids* 1969 **12** (11), 2418–2425.

79. Morton K. W. Finite amplitude compression waves in a collision-free plasma. *Phys. Fluids* 1967, **7** (11), 1800–1815.

80. Mukanova B. G. Studying a viscous film motion along a sloping plane (in Russian). ITPM SO AN SSSR, Novosibirsk, 1981.

81. Nemirovsky S. K. Non-linear waves of the second sound in He II in the vicinity of T_λ (in Russian). IT SO AN SSSR, Novosibirsk, 1976, Preprint 08-76.

82. Nesterikhin Yu. E., Ponomarenko A. G. and Yablotchnikov B. A. On generation of collisionless shock waves propagating

along a magnetic field (in Russian). *Pis'ma v ZETF* 1966, **4** (1), 10–15.

83. Orszag S. A.   Numerical simulation of incompressible flows within simple boundaries. Galerkin (spectral) representations. *Study Appl. Math.* 1971, **50** (4), 293–327.

84. Pandharipande V. R.   Hyperonic matter. *Nucl. Phys.* 1971, **178** (1), 123–144.

85. Paul J. W.   Review of experimental studies of collisionless shocks propagating perpendicular to a magnetic field. In: *Collision-Free Shocks in the Laboratory and Space*. Frascati, 1969, 97–122.

86. Petviashvili V. I.   Non-one-dimensional solitons (in Russian). In: *Nonlinear Waves*. Nauka, Moscow, 1979, 5–19.

87. Popov Yu. P. and Samarsky A. A.   Completely conservative difference schemes for magnetohydrodynamic equations (in Russian). *Zh. Vytch. Mat. i Mat. Fiz.* 1978, **10** (4), 990–998.

88. Roberts K. and Potter D. Magnetohydrodynamic calculations. In: *Methods in Computational Physics*, ed B. Alder, S. Fernbach, M. Rotenberg. Academic Press, New York, 1970, vol. 9.

89. Roberts K. V. and Taylor I. B.   Magnetohydrodynamic equations for a finite Larmor radius. *Phys. Rev. Lett.* 1962, **8** (5), 197–198.

90. Robson A. E.   Experiments on oblique shock waves. In: *Collision-Free Shocks in the Laboratory and Space*. Frascati, 1969, 159–176.

91. Roskes G. J.   Three-dimensional long waves on a liquid film. *Phys. Fluids* 1970, **13** (6) 1440–1445.

92. Rozhdestvensky B. L. and Yanenko N. N.   Systems of quasilinear equations. *Am. Math. Soc.* Monograph 55 (1983).

93. Sagdeev R. Z.   Collective processes and shock waves in a rarefied plasma. In: *Reviews of Plasma Physics*, ed. M. A. Leonotich. Consultants' Bureau, New York, 1966, vol. 4, p. 51.

94. Sagdeev R. Z.   On Ohm's law in unstable plasmas. In: *Proc. of the Symp. on Appl. Math.* New York, 1967, vol. 18, 281–289.

95. Samarsky A. A.   Numerical methods of low-temperature plas-

ma. In: *Advances in Plasma Physics*, New York, 1974, vol. 5, 185–209.

96. Samarsky A. A. and Popov Yu. P. *Difference Schemes of Gasdynamics* (in Russian). Nauka, Moscow, 1975.

97. Samarsky A. A., Volosevitch P. P., Voltchinskaya M. I. and Kurdyumov S. P. Finite-difference methods for solving one-dimensional unsteady problems of magnetohydrodynamics (in Russian). *Zh. Vytch. Mat. i Mat. Fiz.* 1968, **8** (5), 1025–1038.

98. Shapiro V. D. and Shevtchenko V. I. Quasi-linear theory of instability of the plasma with anisotropic ion velocity distribution (in Russian). *Zh. Eksp. i Tekh. Fiz.* 1963, **45** (5) (11), 1612–1624.

99. Sigov Yu. S. and Khodyrev Yu. V. On the theory of discrete models of a rarefied plasma (in Russian). *Num. Meth. of Fluid Mech.*, Novosibirsk, 1976, **7** (2), 109–117.

100. Sgro A. G. and Nielson S. W. Hybrid model studies of ion dynamics and magnetic field diffusion during pinch implosions. *Phys. Fluids* 1976, **19** (1), 126–133.

101. Stevenson J. C. Numerical studies of magnetic field annihilation. *J. Plasma Phys.* 1972, **7** (2), 293–311.

102. Tikhonov A. N., Samarsky A. A., Zaklyaz'minsky L. A. *et al.* Non-linear effects of formation of a self-supporting high-temperature gas layer in unsteady processes of magnetohydrodynamics (in Russian). *Dokl. AN SSSR* 1967, **173** (4), 80–83.

103. Tsvelodub O. Yu. Stationary plane waves on a falling liquid film (in Russian). In: *Thermophysical Investigations*. IT SO AN SSSR, Novosibirsk, 1977.

104. Tverskoy B. A. On structure of shock waves in plasma (in Russian). *Zh. Eksp. i Teor. Fiz* **46** (5), 1653–1663.

105. Ugai M. and Tsuda T. Magnetic fluid-line reconnection by localized enchancement of resistivity. *J. Plasma Phys.* 1977, **18** (2), 451–471.

106. Vliegenthaart A. C. On finite-difference methods for the Korteweg–de Vries equation. *J. Eng. Math.* 1971, **5** (2), 137–155.

107. Vol'mir A. S. *Problems of Hydroelasticity* (in Russian). Nauka, Moscow, 1979.

108. Washimi H. and Taniuti T. Propagation of ion-acoustic solitary waves of small amplitudes. *Phys. Rev. Lett.* 1966, **17** (17), 966–971.

109. Wijngaarden L. On the equations of motion for mixtures of liquid and gas bubbles. *J. Fluid Mech.* 1965, **33** (2), 465–474.

110. Yanenko N. N. *The Method of Fractional Steps*. Springer, New York, 1971.

111. Zabusky N. J. and Kruskal M. D. Interaction of solitons in a collisionless plasma and the recurrence of initial states. *Phys. Rev. Lett.* 1965, **15** (6), 240–243.

112. Zaslavsky G. M. and Moiseev S. S. On effects of magnetic viscosity on stability of a plasma with anisotropic pressure (in Russian). *Prikl. Mekh. i Tekh. Fiz.* 1962, (6), 119–120.